7th International Workshop on New Computational Methods for Inverse Problems (NCMIP 2017)

AA006822

Journal of Physics: Conference Series Volume 904

Paris-Saclay, France
12 May 2017

ISBN: 978-1-5108-5035-4
ISSN: 1742-6588

Printed from e-media with permission by:

Curran Associates, Inc.
57 Morehouse Lane
Red Hook, NY 12571

CURRAN ASSOCIATES INC.
proceedings
.com

Some format issues inherent in the e-media version may also appear in this print version.

Printed by Curran Associates, Inc. (2017)

Institute of Physics
Dirac House, Temple Back
Bristol BS1 6BE UK

Phone: 44 1 17 929 7481
Fax: 44 1 17 920 0979

techtracking@iop.org

Additional copies of this publication are available from:

Curran Associates, Inc.
57 Morehouse Lane
Red Hook, NY 12571 USA
Phone: 845-758-0400
Fax: 845-758-2633
Email: curran@proceedings.com
Web: www.proceedings.com

Table of contents

Volume 904

7th International Conference on New Computational Methods for Inverse Problems
12 May 2017, Paris-Saclay, France

Accepted papers received: 6 October 2017
Published online: 22 October 2017

Preface

Papers

Foreword

This volume of Journal of Physics: Conference Series is dedicated to the scientific contributions presented during the 7th International Conference on New Computational Methods for Inverse Problems, NCMIP 2017 (http://complement.farman.ens-cachan.fr/NCMIP_2017.html). This conference took place at Ecole normale supérieure Paris-Saclay, on May 12, 2017. The prior editions of NCMIP also took place in Cachan, France, firstly within the scope of ValueTools Conference, in May 2011, and secondly at the initiative of Institut Farman, in May 2012, May 2013, May 2014, May 2015 and May 2016.

The New Computational Methods for Inverse Problems (NCMIP) conference focused on recent advances in the resolution of inverse problems. Indeed, inverse problems appear in numerous scientific areas such as geophysics, biological and medical imaging, material and structure characterization, electrical, mechanical and civil engineering, and finances. The resolution of inverse problems consists in estimating the parameters of the observed system or structure from data collected by an instrumental sensing or imaging device. Its success firstly requires the collection of relevant observation data. It also requires accurate models describing the physical interactions between the instrumental device and the observed system, as well as the intrinsic properties of the solution itself. Finally, it requires the design of robust, accurate and efficient inversion algorithms. Advanced sensor arrays and imaging devices provide high rate and high volume data; in this context, the efficient resolution of the inverse problem requires the joint development of new models and inversion methods, taking computational and implementation aspects into account. During this one-day conference, researchers had the opportunity to bring to light and share new techniques and results in the field of inverse problems.

The topics of the conference were: algorithms and computational aspects of inversion, Bayesian estimation, Kernel methods, learning methods, convex optimization, free discontinuity problems, metamodels, proper orthogonal decomposition, reduced models for the inversion, non-linear inverse scattering, image reconstruction and restoration, and applications (bio-medical imaging, non-destructive evaluation...).

NCMIP 2017 was a one-day conference held in May 2017 which attracted around seventy attendees. Each of the submitted papers has been reviewed by two reviewers. There have been nineteen accepted papers. In addition, three international speakers were invited to present a longer talk.

The conference was supported by Institut Farman (ENS Paris-Saclay, CNRS) and endorsed by the following French research networks: GDR ISIS, GDR MIA, GDR MOA, GDR Ondes. It also was supported by Digiteo. The program committee acknowledges the following research laboratories: CMLA, LMT, LURPA and SATIE.

Eric Vourc'h and Thomas Rodet

Conference co-chairs:

Eric Vourc'h, SATIE laboratory, Ecole Normale Supérieure de Cachan, CNRS, France
Thomas Rodet, SATIE laboratory, Ecole Normale Supérieure de Cachan, CNRS, France

Technical program committee:

Alexandre Baussard, ENSTA Bretagne, Lab-STICC, France
Marc Bonnet, ENSTA, ParisTech, France
Laure Blanc-Féraud, I3S laboratory and INRIA Nice Sophia-Antipolis, France
Antonin Chambolle, CMAP, Ecole Polytechnique, CNRS, France
Oliver Dorn, School of Mathematics, University of Manchester, UK
Cécile Durieu, SATIE, ENS Paris-Saclay, CNRS, France
Laurent Fribourg, LSV, ENS Paris-Saclay, CNRS, France
Jerôme Idier, IRCCyN Laboratory, Ecole Centrale de Nantes, France
Pierre-Yves Joubert, IEF, Paris-Sud University, CNRS, France
Marc Lambert, Geeps Laboratory, CNRS, CentraleSupElec, Paris-Sud University, France
Giacomo Oliveri, eledia research center/eledia@L2S group, University of Trento, Italy
Dominique Lesselier, L2S Laboratory, CNRS, CentraleSupElec, Paris-Sud University, France
Matteo Pastorino, DIBE, University of Genoa, Italy
Gabriel Peyré, Ceremade laboratory, University of Paris Dauphine, France
Anthony Quinn, Trinity College Dublin, Ireland
Christian Rey, Safran Tech, France
Marco Salucci, eledia research center/eledia@L2S group, University of Trento, Italy
Jean-Yves Tourneret, IRIT laboratory, ENSEEIHT (Toulouse), France

Local chairs:

Virginie Pauchont, CMLA Laboratory, ENS Paris-Saclay, Paris-Sud University, CNRS, France
Sophie Abriet, SATIE Laboratory, ENS Paris-Saclay, Paris-Sud University, CNRS, France
Béatrice Bacquet, SATIE Laboratory, ENS Paris-Saclay, Paris-Sud University, CNRS, France
Dominique Bach, SATIE Laboratory, ENS Paris-Saclay, Paris-Sud University, CNRS, France

Invited speakers

Thomas Pock, Institute of Computer Graphics and Vision (ICG), TU, Graz, Austria
Silvia Bonettini, Università degli studi di Ferrara, dipartimento di matematica informatica, Ferrara, Italy.
Marion Darbas, LAMFA, CNRS UMR 7352, Université de Picardie Jules Verne, France.

Peer review statement

All papers published in this volume of *Journal of Physics: Conference Series* have been peer reviewed through processes administered by the proceedings Editors. Reviews were conducted by expert referees to the professional and scientific standards expected of a proceedings journal published by IOP Publishing.

IOP Conf. Series: Journal of Physics: Conf. Series **904** (2017) 012001 doi:10.1088/1742-6596/904/1/012001

Vascular blood flow reconstruction from tomographic projections with the adjoint method and receding optimal control strategy

B Sixou[1], L.Boissel[1], M.Sigovan[1]
[1] CREATIS, INSA-Lyon; Inserm, U1044; CNRS UMR 5220; Université Lyon 1; Université de Lyon; 69621, Lyon, France
E-mail: `bruno.sixou@insa-lyon.fr`

Abstract. In this work, we study the measurement of blood velocity with contrast enhanced computed tomography. The inverse problem is formulated as an optimal control problem with the transport equation as constraint. The velocity field is reconstructed with a receding optimal control strategy and the adjoint method. The convergence of the method is fast.

1. Introduction

The blood velocity in arteries is an important clinical parameter [1]. Recently, we have proposed a method to evaluate this velocity with contrast agent enhanced tomography [2]. In this work, the inverse problem of contrast enhanced computerized tomography (CT) reconstruction of vascular blood flow is formulated as an optimization problem with partial differential equation constraints. The reconstruction is based on the acquisition of 2D CT projections with a projection angle perpendicular to main flow propagation direction and increasing linearly with time. The inversion is regularized with the partial differential equation constraint describing the transport of the contrast agent coupling its density $f(\mathbf{x}, t)$ and the flow field $\mathbf{w}(\mathbf{x}, t)$. In this former approach, the pde constraint is used as a side constraint in the optimization problem and the variables $(f, \mathbf{w})$ are treated as independant optimization variables. Yet, the efficient solution of the optimality system is a very challenging task. Good results have been obtained for the density but not for the velocity field. Moreover, the computing time is very long. In this new work, a suboptimal receding optimal control stategy is used to get a simpler optimization problem and to reduce the computing time. This type of approach consists in replacing the initial optimization problem by a sequence of optimization problems on smaller time intervals. It has been used for the optimal control of fluid flows [3, 4]. Moreover, we consider here a new approach for solving the optimization problem. The pde constraint is eliminated with a control to state operator $f = f(\mathbf{w})$, and the only optimization variable is the fluid velocity $\mathbf{w}$. A gradient type method is used to minimize the reduced functional. The efficient evaluation of the derivative uses the adjoint method [5]. This paper is organized as follows. In Section 1, the dynamic tomography problem considered is summarized and it is formulated as an optimal control problem. The receding optimal control strategy and the adjoint method are presented in the second section. We display numerical results with a simple synthetic fluid flow and we show that this new formulation improves the reconstruction of the flow field.

2. The contrast enhanced tomography inverse problem

2.1. Optimal control formulation

We summarize here the basic optimal control formulation presented in [2] which is the starting point of our new approach. The phantom considered and the CT acquisition geometry are shown in Figure 1. This phantom simulates a straight arterial vessel segment oriented along the z axis. During the propagation into the phantom, two dimensional projections are acquired with a projection direction perperpendicular to the z axis and a projection angle $\theta(t)$ increasing linearly with time t.

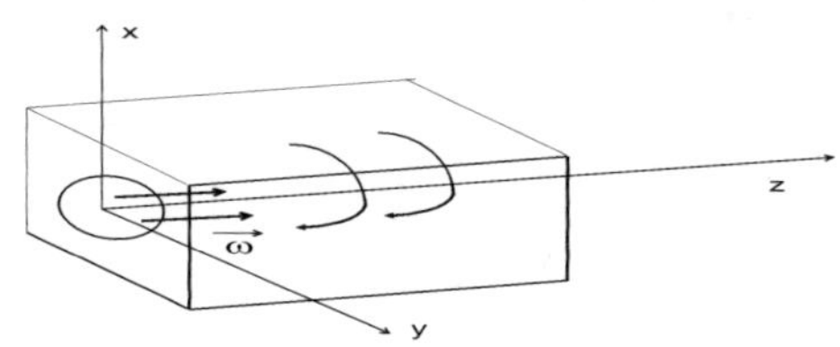

Figure 1. Flow phantom and CT acquisition geometry.

.

The bounded spatial domain is dentoted Ω and the density of the contrast agent to be reconstructed $f(\mathbf{x}, t)$. The space-time box for the problem is denoted as $Q = \Omega \times [0, T]$, where $[0, T]$ is the time domain. The structures other than the contrast material are assumed to have a zero density. We assume the contrast agent follows a scalar diffusion-transport equation with a velocity field, $\mathbf{w} = (u, v, w) : [0, T] \times \Omega \to \mathbb{R}^3$ and diffusion coefficient D. It is injected on a disk $(\mathcal{D})$ in the plane $z = 0$.

$$\begin{cases} E(f, \mathbf{w}) = \frac{\partial f(\mathbf{x},t)}{\partial t} + u\frac{\partial f(\mathbf{x},t)}{\partial x} + v\frac{\partial f(\mathbf{x},t)}{\partial y} + w\frac{\partial f(\mathbf{x},t)}{\partial z} - D\triangle f(\mathbf{x}, t) = 0 \quad \forall t \in [0, T] \quad \forall \mathbf{x} \in \Omega \\ f(\mathbf{x}, t) = 1 \;\; on \;\; (\mathcal{D}) \;\; \forall t \in [0, T] \quad f(\mathbf{x}, 0) = 0 \;\; on \;\; \Omega - (\mathcal{D}) \end{cases} \tag{1}$$

Compared to our former work [2], the addition of a diffusion term ensures the unicity of f for a smooth field $\mathbf{w}$. For the sake of simplicity, we have suppressed the incompressibility constraint. Let $X = L_2(\Omega)$, our aim is to find $f : [0, T] \to X$ from the linear equations, $R_{\theta(t)} f(u_1, u_2) = p^{\delta}_{\theta(t)}$, where $p^{\delta}_{\theta(t)}$ are the noisy measured projections, $R_{\theta(t)}$ is the linear Radon operator for the projection angles $\theta(t)$, and u_1 and u_2 are the detector coordinates. The cost functional considered is:

$$J(f, \mathbf{w}) = \frac{1}{2} \int_0^T \|R_{\theta(t)} f(u_1, u_2) - p^{\delta}_{\theta(t)}(t)\|_2^2 dt + \frac{\alpha}{2} \int_0^T \|\nabla \mathbf{w}\|_2^2 dt \tag{2}$$

where $\|\nabla \mathbf{w}\|_2$ the Frobenius norm of the Jacobian matrix of the flow field $\mathbf{w}$ and α a regularization parameter. In our former work, we considered the following optimization problem, $\min_{f,\mathbf{w}} J(f, \mathbf{w})$ with $E(f, \mathbf{w}) = 0$.

2.2. The receding horizon technique and the new optimal control formulation

Solving the time dependent optimal control problem governed by the former PDE requires a significant computing time and a large storage power. Several suboptimal solutions have been proposed [3, 4] to make large problems feasible and to reduce the computing time. We consider here the receding horizon control method. The idea of the receding horizon control is to replace the optimal control problem on the full time horizon by a succession of simpler optimal

IOP Conf. Series: Journal of Physics: Conf. Series **904** (2017) 012001 doi:10.1088/1742-6596/904/1/012001

control problems on short control horizons. Let $[0, T]$ denote the time interval over which the optimization is performed. We chose time values $0 = T_0 < ..T_k.. < T_K = T$ to partition this time interval. On each time sub-interval, we assume that we have constant velocity fields $\mathbf{w}_k$. For k=0.....K-1, we minimize sequentially the functionals

$$\hat{J}_k(\mathbf{w}_k) = J_k(f(\mathbf{w}_k), \mathbf{w}_k) \tag{3}$$

with

$$J_k(f, \mathbf{w}) = \frac{1}{2}\int_{T_k}^{T_{k+1}} \|R_{\theta(t)}f(u_1, u_2) - p^{\delta}_{\theta(t)}(t)\|^2 dt + \frac{\alpha}{2}\int_{T_k}^{T_{k+1}} \|\nabla \mathbf{w}\|_2^2 dt = \int_{T_k}^{T_{k+1}} H(f, \mathbf{w})dt \tag{4}$$

subject to the partial differential equation (1) over $[T_k, T_{k+1}] \times \Omega$. On each of the intervals $[T_k, T_{k+1}]$ we have to consider a vector $\mathbf{q}$ of unknown parameters $\mathbf{q} = [a \quad \mathbf{w}]$ with $f(T_k, .) = a$ and $a \in L_2(\Omega)$, corresponding to the initial condition. The initial condition can be written $g(f(T_k), \mathbf{q}) = f(T_k) - a = 0$.

2.3. The adjoint method

In our first approach [2], the variables $(f, \mathbf{w})$ are both considered as optimization variables. We present here an alternative approach. We consider here a parabolic equation (1) with a diffusion term which admits a unique solution for sufficiently smooth $\mathbf{w}$ [6, 7]. The former optimal control problem is equivalent to the reduced optimization problem where only the control variable $\mathbf{w}$ appears:

$$\min \hat{J}(\mathbf{w}) = J(f(\mathbf{w}), \mathbf{w}) \tag{5}$$

where $f(\mathbf{w})$ satisfie $E(f(\mathbf{w}), \dot{f}(\mathbf{w}), \mathbf{w}) = 0$. To derive the optimality conditions and to calculate the gradient of the reduced functional $\hat{J}_k$, we have to introduce the Lagrangian corresponding to the optimisation problem:

$$L_k(f, \mathbf{q}) = J(f(\mathbf{w}_k), \mathbf{w}_k) + \int_{T_k}^{T_{k+1}} < Ef(t), p_k(t) > dt + < \mu_k, g(f(T_k), \mathbf{q}_k) > \tag{6}$$

We consider two dual variables $p_k : [T_k, T_{k+1}] \times \Omega \to \mathbb{R}$ and $\mu_k : [T_k, T_{k+1}] \times \Omega \to \mathbb{R}$ for the constraints. In order to apply the adjoint method, we compute the directional derivative of L_k with respect to $\mathbf{q}_k$ in the direction $\mathbf{r}$:

$$\begin{aligned} L_{k,\mathbf{q}_k}(f, \mathbf{q}_k)\mathbf{r} = \int_{T_k}^{T_{k+1}} & [\partial_f H d_{\mathbf{q}_k} f\mathbf{r} + \partial_{\mathbf{q}_k} H\mathbf{r}]dt + < p_k, \partial_f E d_{\mathbf{q}_k} f\mathbf{r} + \partial_{\dot{f}} E d_{\mathbf{q}_k} \dot{f}\mathbf{r} \\ & + \partial_{\mathbf{q}_k} E\mathbf{r} > dt + < \mu, \partial_{f(T_k)} g d_{\mathbf{q}_k} f(T_k)\mathbf{r} + \partial_{\mathbf{q}_k} g\mathbf{r} > \end{aligned} \tag{7}$$

With an integration by part,

$$\begin{aligned} \int_{T_k}^{T_{k+1}} < p_k, \partial_{\dot{f}} E d_{\mathbf{q}_k} \dot{f}\mathbf{r} > dt = & < p_k, \partial_{\dot{f}} E d_{\mathbf{q}_k} f\mathbf{r} > |_{T_k}^{T_{k+1}} \\ & - \int_{T_k}^{T_{k+1}} [< \dot{p}_k, \partial_{\dot{f}} E > + < p_k, d_t \partial_{\dot{f}} E d_{\mathbf{q}_k} f]\mathbf{r} dt \end{aligned} \tag{8}$$

Substituting in (10), we obtain:

$$L_{k,\mathbf{q}_k}(f, \mathbf{q}_k)\mathbf{r} = \int_{T_k}^{T_{k+1}} [\partial_f H d_{\mathbf{q}_k} f\mathbf{r} + \partial_{\mathbf{q}_k} H\mathbf{r}] + < p_k, (\partial_f E - d_t \partial_{\dot{f}} E d_{\mathbf{q}_k} f)\mathbf{r} > \tag{9}$$

$$- < \dot{p}_k, \partial_{\dot{f}} E d_{\mathbf{q}_k} f_k \mathbf{r} >) + f_{\mathbf{q}_k}\mathbf{r} + < p_k, \partial_{\mathbf{q}_k} E\mathbf{r} >]dt \tag{10}$$

$$- < p_k, \partial_{\dot{f}} E d_{\mathbf{q}_k} f\mathbf{r}|_{T_{k+1}} - < p(T_k), \partial_{\dot{f}} E|_{T_k} d_{\mathbf{q}_k} f(T_k)\mathbf{r} > + < \mu, g_{f(T_k)}|_{T_k} d_{\mathbf{q}_k} f(T_k)\mathbf{r} > + < \mu, g_{\mathbf{q}_k}\mathbf{r} > \tag{11}$$

IOP Conf. Series: Journal of Physics: Conf. Series **904** (2017) 012001 doi:10.1088/1742-6596/904/1/012001

The principle of the adjoint method is to avoid calculating the total derivative $d_{\mathbf{q}_k} f$ by setting:

$$\partial_f H + (\partial_f E - d_t \partial_{\dot{f}} E)^* p_k - \partial_{\dot{f}} E^* \dot{p_k} = 0 \tag{12}$$

For the problem considered,

$$\partial_f H = R^t_{\theta(t)}(R_{\theta(t)} f(\mathbf{x}, t) - p^\delta_{\theta(t)}(t)) \quad \partial_{\dot{f}} E = I \quad \partial_f E = \mathbf{w}_k.\nabla - D\triangle \tag{13}$$

We obtain the following adjoint partial differential equation:

$$R^t_{\theta(t)}(R_{\theta(t)} f(\mathbf{x}, t) - p^\delta_{\theta(t)}(t)) + (-\mathbf{w}_k.\nabla - D\triangle)p_k - \dot{p_k} = 0 \quad p_k(T_{k+1}) = 0 \tag{14}$$

Similarly the optimality with respect to $f(T_k)$ gives $\partial_{\dot{f}} E|^*_{T_k} p_k(T_k) = g_{f(T_k)}|^*_{T_k} \mu$ or

$$\mu = g^{-*}_{f(T_k)} \partial_{\dot{f}_k} E|^*_{T_k} p_k(T_k) = -p_k(T_k) \tag{15}$$

The gradient of the Lagrangian is then given by:

$$\nabla \hat{J}'(\mathbf{q}_k) = \nabla L_{\mathbf{q}_k}(\mathbf{q}_k) = \int_{T_k}^{T_{k+1}} (H_{\mathbf{q}_k} + \partial_{\mathbf{q}_k} E^* p_k) dt + g^*_{\mathbf{q}_k} g^{-*}_{f(T_k)} \partial_{\dot{f}} E|^*_{T_k} p_k(T_k) \tag{16}$$

The first term $\int_{T_k}^{T_{k+1}}(-\alpha\triangle\mathbf{w}_k + p_k\nabla f)dt$ corresponds to the gradient for the $\mathbf{w}$ component of the vector $\mathbf{q}_k$ and the second one, $-p_k(T_k)$ to the initial value component $f(T_k)$. The evaluation of the gradient $\nabla\hat{J}(\mathbf{q}_k)$ for a given control function $\mathbf{w}_k$ can be summarized in the algorithm:
1) Solve the forward flow equation to obtain $f(\mathbf{x}, t)$ from the current approximation $\mathbf{w}_k$(Eq.1).
2) Solve the backward adjoint equation to obtain the adjoint variable p_k(Eq.14).
3) Return the gradient, $(\int_{T_k}^{T_{k+1}}(-\alpha\triangle\mathbf{w}_k + p_k\nabla f)dt, -p_k(T_k))$

3. Numerical algorithm

3.1. Simulation details

The values of the components (u, v, w) of the flow field $\mathbf{w}$ and of the density f are discretized on a grid of size $N_s = 53 \times 53 \times 53$ for the spatial domain Ω and on a grid of size $N_t = 100$ for the time domain $[0, T]$. For each time t_k in the time domain $[0, T]$, a projection angle $\theta_k = 2\pi t_k/T$ is used for the Radon projection operator. During the CT gantry rotation, 100 parallel projections are thus acquired. Moreover, 77 projections rays are acquired for each value of the projection angle. The ground truth solution f^* is obtained from the scalar transport equation with a uniform velocity along the z axis, slowly decreasing with time, $\mathbf{w}^* = (u^*, v^*, w^*) = (0, 0, 10 + 5exp(-t/10))$. The step size for the receding control strategy $T_{k+1} - T_k$ is 6 five times the time step used for the projection $t_{k+1} - t_k$. The total horizon time is thus divided in 16 subintervals. The initial density of the contrast agent is $f(\mathbf{x}, z = 0, t = 0) = 1$ for the pixels inside a disk of diameter 17 around the z axis. A Gaussian white noise with peak-to-peak signal-to-noise ratio (PPSNR) of 48 dB was added to the projections. The regularization parameter value $\alpha = 2.10^{-1}$ is chosen to have the best decrease of the discrepancy term. For the integration of the partial differential equation, the problem is discretized by finite difference and we use an explicit scheme of order 2 with numerical diffusion. For this numerical scheme, the CFL condition is verified, $CFL = \|\mathbf{w}\|dt/dx < 1$. For the simulation of the transport equation, the time step $dt = 0.005$ is used.The adjoint equation is solved by the same method as described before by reversing time. The reduced functional min $\hat{J}$ is minimized iteratively with a steepest descent method on each subinterval with an inner loop sing the negative of the gradient of $\hat{J}$ as search direction. The step length ρ_k is chosen as, $\rho_k = \arg\min_\rho \hat{J}(\mathbf{w}_k + \rho\nabla\hat{J}(\mathbf{w}_k))$.

3.2. Results

Sections of the reconstructed w and of the contrast agent density obtained at the end of the optimization process for $z = 5$ and $t = 10s$ are displayed in Figure 2. In order to validate the method, on each subinterval $[T_k, T_{k+1}]$, we have studied the evolution of the L_2 error term for the velocity field $\mathbf{w}_k^j$ and the reconstructed density f_k^j with the inner iterations j during the steepest descent:

$$D(f_k^j) = \sum_{t_i=T_k}^{T_{k+1}} \|f_k^j(\mathbf{x}, t_i) - f^*(\mathbf{x}, t_i)\|^2 \tag{17}$$

$$D(\mathbf{w}_k^j) = \sum_{t_i=T_k}^{T_{k+1}} \|u^*(\mathbf{x}, t_i) - u_k^j(\mathbf{x}, t_i)\|_2^2 + \|v^*(\mathbf{x}, t_i) - v_k^j(\mathbf{x}, t_i)\|_2^2 + \|w^*(\mathbf{x}, t_i) - w_k^j(\mathbf{x}, t_i)\|_2^2 \tag{18}$$

where the sum is extended to the times $t_i \in [T_k, T_{k+1}]$. Some error curves are presented on Figure 3 for a selected time interval $[T_k, T_{k+1}] = [6, 12]$. A large and fast reduction of the error is obtained for the reconstructed density and the velocity field. We have also calculated the total error terms on the whole horizon time $[0, T]$ as a function of the total number of outer iterations. The decreases for the reconstruction errors for the new method are compared to the ones of our former approach [2] in Figure 4. The simulation takes 800 s for the case investigated for the receding horizon technique and 20000 for the former method. The new approach presented in this work clearly outperforms the former in terms of speed for similar reconstruction results.

4. Conclusion

In this work we have presented a new method for the reconstruction of the flow velocity of a contrast agent from CT measurements with projections perpendicular to the main axis of propagation. The propagation equation of the contrast agent is used to formulate an unconstrained optimization problem where the unknown is the velocity field. The optimal control problem is solved with a receding horizon technique and with the adjoint method. This new method is tested on a simple phantom. It is much faster than our former approach where the full time interval is considered and where the velocity field and the density field are both considered as optimization variables. In the following, we will investigate second order optimization methods on more complex phantoms and velocity fields.

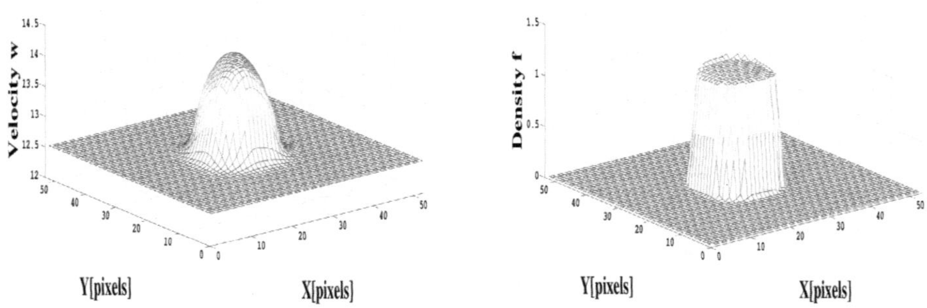

Figure 2. Sections of the reconstructed velocity and density for z=5, t=10s at the end of the optimization process.

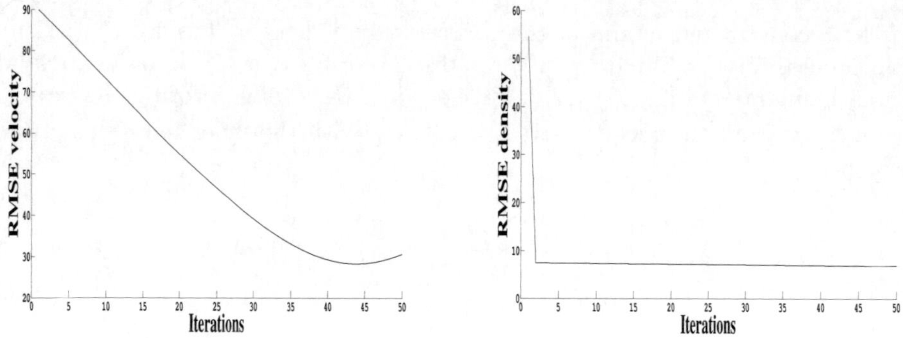

Figure 3. Evolution with the iterations of the RMSE for the velocity and the density on a sub-interval $[T_k, T_{k+1}] = [6, 12]$.

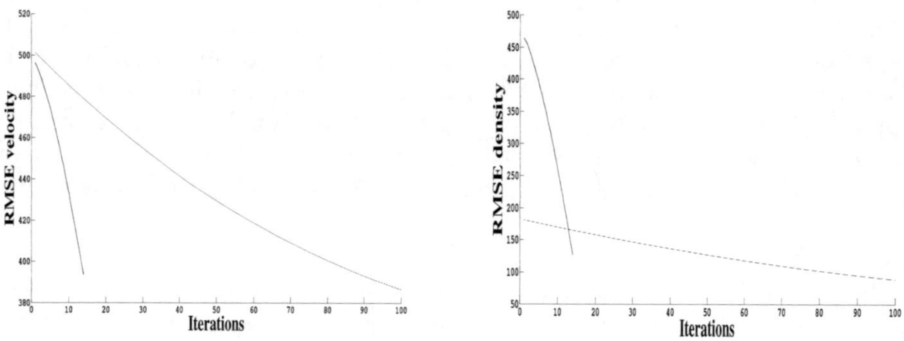

Figure 4. Evolution with the iterations of the RMSE for the velocity and the density. Black curve: receeding horizon technique and adjoint method, blue curve: optimal control strategy detailed in [2].

References

[1] S.Prevrhal, C.H.Forsythe, R.J.Harnish, M.Saeed, B.M.Yeh "CT angiographic measurement of vascular blood fowl velocity by using projection data", *Radiology*, vol.261, 923-929 , 2011.

[2] B.Sixou, L.Boussel and M.Sigovan "Contrast enhanced computerized tomography measurement of vascular blood floww", *NCMIP2016*

[3] M.Hinze and K.Kunisch "Three control methods for time-dependent fluid flow", *Flow, turbulence and combustion* , vol.65, 273:298 , 2000.

[4] M.Choi "Suboptimal control of turbulent flow using control theory", *Proceedings of the International Symposium on Mathematical Modelling of Turbulent Flows* , Tokyo, Japan, 1995.

[5] M.Herzog and K.Kunisch "Algorithms for pde-constrained optimization", *GAMM-Mitteilungen*, vol.33, 163-176 , 2010.

[6] O.A.Ladyzhenskaja, V.A.Solonnikov and N.N.Uraltzeva, " Linear and quasilinear equations of parabolic type ", American Math.Soc., Providence, RI, 1968.

[7] A.Friedman "Parabolic differential equations of parabolic type", Dover Books of Mathematics, 1964.

[8] L.C.Evans Partial differential equations, *American Math.Soc. Providence, RI*, 1998.

IOP Conf. Series: Journal of Physics: Conf. Series **904** (2017) 012002 doi:10.1088/1742-6596/904/1/012002

A compressive sensing-based computational method for the inversion of wide-band ground penetrating radar data

A Gelmini[1], G Gottardi[1] and T Moriyama[2]

[1] ELEDIA Research Center (ELEDIA@UniTN - University of Trento)
Via Sommarive 9, I-38123 Trento, Italy
[2] ELEDIA Research Center (ELEDIA@UniNAGA - University of Nagasaki)
852-8521, Nagasaki, Japan
E-mail: {angelo.gelmini, giorgio.gottardi, toshifumi.moriyama}@eledia.org

Abstract. This work presents an innovative computational approach for the inversion of wide-band ground penetrating radar (*GPR*) data. The retrieval of the dielectric characteristics of sparse scatterers buried in a lossy soil is performed by combining a multi-task Bayesian compressive sensing (*MT-BCS*) solver and a frequency hopping (*FH*) strategy. The developed methodology is able to benefit from the regularization capabilities of the *MT-BCS* as well as to exploit the multi-chromatic informative content of *GPR* measurements. A set of numerical results is reported in order to assess the effectiveness of the proposed *GPR* inverse scattering technique, as well as to compare it to a simpler single-task implementation.

1. Introduction

During the last decades, ground penetrating radar (*GPR*) gained particular attention as a valid prospecting technology in many applicative scenarios dealing with the non-invasive investigation of buried domains, such as, for example, civil engineering, military operations, cultural heritage monitoring, and archaeology [1]-[4]. Thanks to such a success, several inverse scattering methodologies, both deterministic [5]-[8] and stochastic [9],[10], have been proposed to retrieve accurate and easy-to-interpret images of what lies below the interface from *GPR* measurements.

Differently from other prospecting tools, a key feature that characterizes *GPR* is the availability of wide-band data. As a matter of fact, such a frequency diversity can be seen as a powerful source of additional information in order to effectively tackle the *ill-posedness* and *non-linearity* issues of the subsurface imaging problem.

Within this framework, different techniques have been proposed in order to process multi-chromatic data. These approaches can be mainly classified in (*i*) multi-frequency (*MF*) [5],[7],[10] and (*ii*) frequency hopping (*FH*) [6] strategies. On the one hand, *MF* techniques are based on the simultaneous processing of the collected wide-band information [10], while on the other hand *FH* ones process each frequency sample in a cascaded manner, starting from the lowest frequency up to the highest one, and exploiting the acquired information at each step in order to initialize the successive one [6]. Given that, even if *MF* strategies could in principle benefit from a larger amount of information to perform the inversion, they must typically solve a more complex problem with a significantly larger number of unknowns with respect to *FH*-based methodologies.

Moreover, it is worth remarking that in many practical scenarios the unknown buried objects are sparse [5]. Given that, compressive sensing (*CS*)-based techniques [11]-[13] could be seen as an interesting candidate in order to exploit such an *a-priori* information about the nature of the sought solution, by further regularizing the *GPR* inverse scattering problem. Following this line of reasoning, this work presents an innovative inversion methodology that effectively integrates a multi-task Bayesian *CS* (*MT-BCS*) solver [11] within a *FH* strategy in order to retrieve accurate reconstructions of the buried scenario. The *MT-BCS* solution at each stage of the frequency hopping loop is initialized in an innovative way, by exploiting the *acquired* information about the detected scatterers at the lowest frequencies. A set of numerical experiments is presented and discussed, in order to assess the effectiveness of the developed *FH-MT-BCS* technique, and to verify its superiority with respect to a single-task *BCS* (*ST-BCS*)-based implementation.

2. *GPR-IS* problem formulation and *FH-MT-BCS* solution approach

Let us consider a two-dimensional scenario consisting of two homogeneous half-spaces separated by a planar interface at $y=0$. The upper half-space is free-space, while the lower half-space is occupied by a lossy medium having complex permittivity $\varepsilon_{b,eq}(f)=\varepsilon_{rb}\varepsilon_0-\mathrm{j}\sigma_b/(2\pi f)$. A buried investigation domain D_{inv} is illuminated by a set of V z-oriented ideal line sources placed at fixed height above the interface and excited by a wideband current signal. Under the hypothesis that one or multiple targets are present in D_{inv}, the electromagnetic interaction between the field radiated by the sources and the half-space scenario generates a measured time-domain total field equal to

$$e_v^{tot}(\mathbf{r}_m,t)=e_v^{scat}(\mathbf{r}_m,t)+e_v^{inc}(\mathbf{r}_m,t);\quad v=1,\dots,V\ ;\quad m=1,\dots,M \tag{1}$$

where $\mathbf{r}_m$ is the location of the *m*-th receiver above the interface ($m=1,\dots,M$, M being the number of probes), while $e_v^{inc}(\mathbf{r}_m,t)$ and $e_v^{scat}(\mathbf{r}_m,t)$ denote the incident and scattered fields, respectively. Once $e_v^{scat}(\mathbf{r}_m,t)$ has been isolated from $e_v^{tot}(\mathbf{r}_m,t)$ [8], it is Fourier-transformed to the frequency domain, and a set of *L* uniformly spaced samples are extracted from the computed *GPR* spectrum within the 3dB bandwidth of the excitation signal ($f_l\in[f_{\min},f_{\max}]$, $l=1,\dots,L$). Accordingly, assuming a contrast source formulation, the following integral equation holds true at each *l*-th frequency

$$E_v^{scat}(\mathbf{r}_m,f_l)=\int_{D_{inv}}G^{ext}(\mathbf{r}_m,\mathbf{r}',f_l)J_v(\mathbf{r}',f_l)d\mathbf{r}';\quad v=1,\dots,V\ ;\quad m=1,\dots,M\ ;\quad l=1,\dots,L \tag{2}$$

where $G^{ext}(\mathbf{r}_m,\mathbf{r}',f_l)$ is the external Green's function for the buried scenario [10], while $J_v(\mathbf{r},f_l)=E_v^{tot}(\mathbf{r},f_l)\tau(\mathbf{r},f_l)$ is the unknown equivalent current modelling the presence of the unknown objects in D_{inv}, $\tau(\mathbf{r},f_l)$ being the contrast function defined at frequency f_l as follows

$$\tau(\mathbf{r},f_l)=(\varepsilon_r(\mathbf{r})-\varepsilon_{rb})-\mathrm{j}\frac{(\sigma(\mathbf{r})-\sigma_b)}{2\pi f_l}\ ;\ l=1,\dots,L. \tag{3}$$

Assuming an additional noise component on the scattered field and partitioning D_{inv} into *N* sub-domains, (2) can be arranged in matrix form as follows

$$\widetilde{\mathbf{E}}_{v,l}^{scat}=\mathbf{G}_l^{ext}\mathbf{J}_{v,l}+\mathbf{n}_{v,l}\ ;\ v=1,\dots,V\ ;\ l=1,\dots,L \tag{4}$$

where $\mathbf{G}_l^{ext}$ is the $M\times N$ Green's matrix, $\widetilde{\mathbf{E}}_{v,l}^{scat}=\left\{\Re\left[\widetilde{E}_v^{scat}(\mathbf{r}_m,f_l)\right],\Im\left[\widetilde{E}_v^{scat}(\mathbf{r}_m,f_l)\right];m=1,\dots,M\right\}$, $\mathbf{J}_{v,l}=\left\{\Re\left[J_v(\mathbf{r}_n,f_l)\right],\Im\left[J_v(\mathbf{r}_n,f_l)\right];n=1,\dots,N\right\}^T$, while the noise vector is $\mathbf{n}_{v,l}=\left\{\Re\left[n_v(\mathbf{r}_m,f_l)\right],\Im\left[n_v(\mathbf{r}_m,f_l)\right];m=1,\dots,M\right\}$. Accordingly, the solution of the inverse problem at the *l*-th frequency step is formulated within the Bayesian framework as follows

$$\hat{\mathbf{J}}_{v,l}=\arg\left\{\max_{\mathbf{J}_{v,l}}P\left(\mathbf{J}_{v,l}\middle|\widetilde{\mathbf{E}}_{v,l}^{scat}\right)\right\}\ ;\ v=1,\dots,V \tag{5}$$

IOP Conf. Series: Journal of Physics: Conf. Series **904** (2017) 012002 doi:10.1088/1742-6596/904/1/012002

where $P\left(\mathbf{J}_{v,l}\middle|\tilde{\mathbf{E}}_{v,l}^{scat}\right)$ is the posterior probability and $\mathbf{J}_{v,l}$ are correlated among the different views ($v = 1,...,V$). The solution of (5) is then found by means of a customized *MT-BCS* solver as

$$\left[\hat{\mathbf{J}}_{v,l}\right]^{opt} = \left[diag(\hat{\mathbf{a}}_l) + \left(\mathbf{G}_l^{ext}\right)^* \mathbf{G}_l^{ext}\right]^{-1} \left(\mathbf{G}_l^{ext}\right)^* \tilde{\mathbf{E}}_{v,l}^{scat} \;\; ; \;\; v = 1,...,V \tag{6}$$

where $(\cdot)^*$ is the transpose conjugate operator and $\hat{\mathbf{a}}_l = \{\hat{a}_{l,n}, n = 1,...,N\}$ is the shared vector of hyper-parameters estimated through a Relevant Vector Machine (*RVM*) solver.
The estimated contrast function $\hat{\tau}(\mathbf{r}_n, f_l)$, $n = 1,...,N$, is then obtained from (6) by averaging over the different views ($v = 1,...,V$) the ratio between retrieved currents and the corresponding retrieved total fields.
Since standard *BCS* implementations do not allow to exploit progressively *acquired* information about the solution (i.e., to use the reconstruction at the l-th frequency to initialize/guide the $(l+1)$-th step), an innovative "constrained" *RVM* solver has been developed as described in the following. A "filtering and clustering" procedure [6] is applied to the retrieved contrast in order to determine the region of interest (*RoI*) $D_l \subset D_{inv}$ ($D_1 = D_{inv}$) in which the buried scatterers have been detected at frequency f_l. Then, the solution at frequency f_{l+1} is obtained though a "constrained" *RVM* by solving (6) limiting the search space only to those entries $\hat{a}_{l+1,n}$ of $\hat{\mathbf{a}}_{l+1}$ that correspond to the $N_l < N$ cells belonging to D_l. More precisely, the *MT-BCS* solution at the f_{l+1} *FH* stage is aimed at retrieving the reduced vector of hyper-parameters

$$\hat{\mathbf{a}}_{l+1} = \{\hat{a}_{l+1,n}; n = 1,..., N_l\} \text{ subject to } \mathbf{r}_n \in D_l . \tag{7}$$

The *FH-MT-BCS* is terminated once one of the following conditions is met: (*i*) the last frequency step has been processed (i.e., $l = L$) or (*ii*) $|A_l - A_{l-1}|/|A_l| \leq \eta$, where A_l and A_{l-1} denote the area of D_l and D_{l-1}, respectively, and η is a suitable threshold.

3. Numerical results

The aim of this section is to assess the performances of the proposed *FH-MT-BCS* algorithm. Towards this end, a set of numerical experiments is reported by generating time-domain synthetic data through the *GprMax2D* simulator [5]. In order to quantify the quality of each reconstruction, the following integral error is computed at a reference frequency $\bar{f}$ arbitrarily chosen within the $[f_{min}, f_{max}]$ range

$$\chi_{tot} = \frac{1}{N}\sum_{n=1}^{N} \frac{\left|\tau(\mathbf{r}, \bar{f}) - \hat{\tau}(\mathbf{r}, \bar{f})\right|}{\left|\tau(\mathbf{r}, \bar{f}) + 1\right|} . \tag{8}$$

The investigation domain D_{inv} is a square of 0.8 m centred at (0,-0.4) m. $V = 20$ z-oriented infinitely long line sources at fixed height above the interface are used to illuminate it, while $M = 19$ probes co-located with the sources are considered at each view to collect the scattered data. Moreover, D_{inv} is filled with a lossy medium with relative permittivity $\varepsilon_{rb} = 4.0$ and conductivity $\sigma_b = 0.001$ S/m. $L = 9$ uniformly spaced frequency samples are extracted from the computed *GPR* spectrum within the $[f_{min}, f_{max}] = [200, 600]$ MHz bandwidth, while $\bar{f} = 400$ MHz.

3.1. Performances against noise

The first set of experiments deals with the retrieval of a buried scatterer composed by two diagonal pixels with relative permittivity $\varepsilon_r^{obj} = 5.0$ and conductivity of $\sigma_b = 0.001$ S/m [Fig. 1(*a*)]. Figure 1 shows the retrieved dielectric profiles by the *FH-MT-BCS* when considering an additive white Gaussian noise on the measured time-domain total field with a signal-to-noise ratio defined as in [6] and equal to $SNR = 60$ dB [Fig. 1(*b*)] and $SNR = 40$ dB [Fig. 1(*d*)].

IOP Conf. Series: Journal of Physics: Conf. Series **904** (2017) 012002 doi:10.1088/1742-6596/904/1/012002

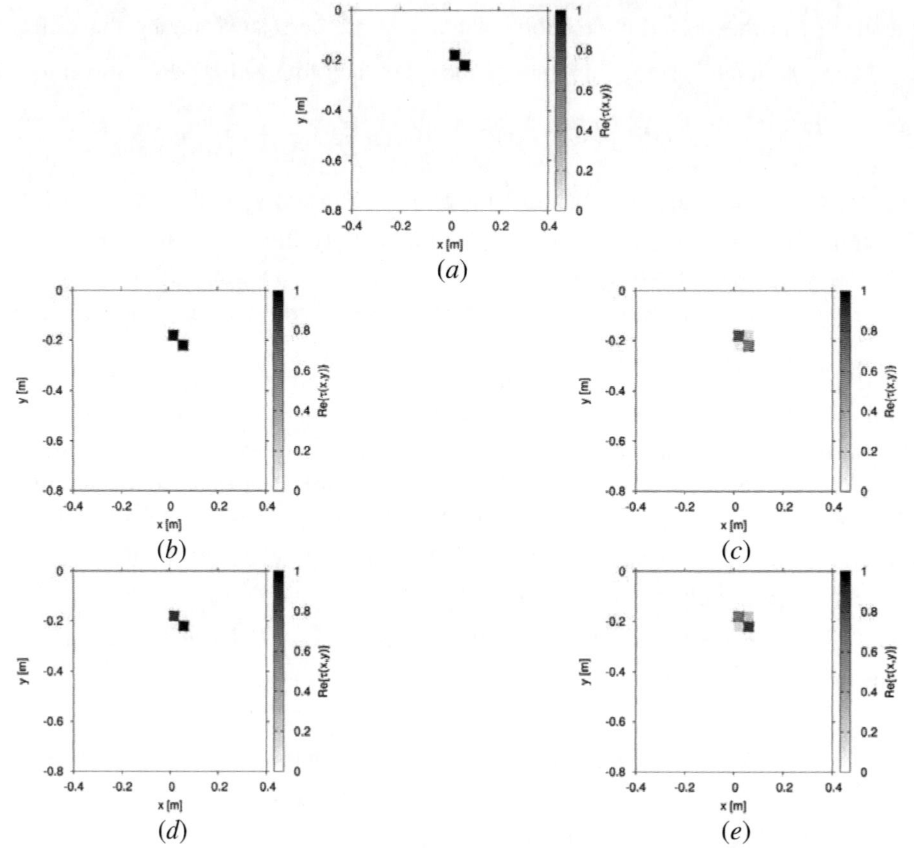

Figure 1. *Numerical Assessment* (*Two diagonal pixels scatterer,* $\varepsilon_r^{obj} = 5.0$, $\sigma^{obj} = 0.001$ S/m) – (*a*) Actual dielectric profile and retrieved solution by (*b*)(*d*) the *FH-MT-BCS* and (*c*)(*e*) the *FH-ST-BCS* methods for (*b*)(*c*) $SNR = 60$ dB and (*d*)(*e*) $SNR = 40$ dB on time-domain total field.

The reconstructions obtained by a single-task *BCS* implementation (*FH-ST-BCS*) are reported, as well [$SNR = 60$ dB - Fig. 1(*c*); $SNR = 40$ dB - Fig. 1(*e*)], this latter not imposing any correlation among the several views in solving (6). As it can be seen, the *FH-MT-BCS* significantly overcomes the *FH-ST-BCS*, as it is also visible by looking at the computed errors vs. the *SNR* reported in Fig. 2.

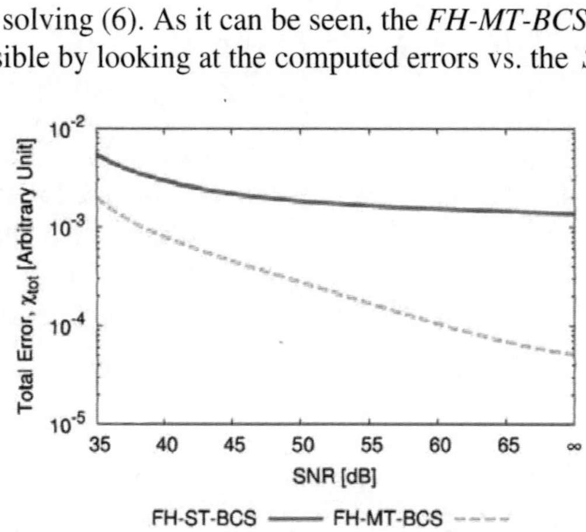

Figure 2. *Numerical Assessment* (*Two diagonal pixels scatterer,* $\varepsilon_r^{obj} = 5.0$, $\sigma^{obj} = 0.001$ S/m) – Behaviour of the reconstruction error versus the *SNR* on time-domain total field.

3.2. *Performances against the relative permittivity of the buried targets*

This second set of numerical benchmarks is aimed at assessing the performances of the *FH-MT-BCS* when considering a variation of the relative permittivity of the unknown buried objects in D_{inv} .

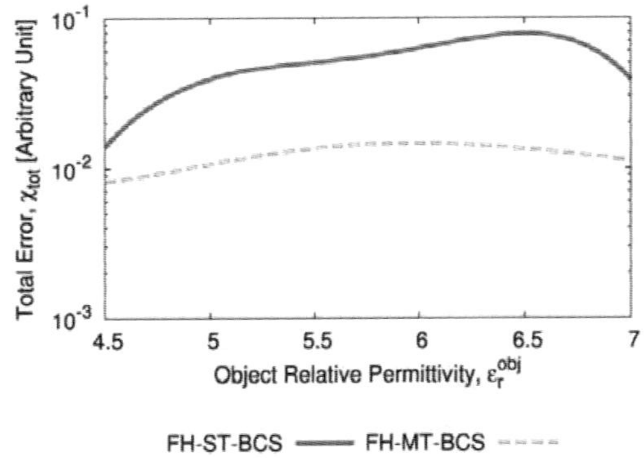

Figure 3. *Numerical Assessment* (*Two bars scatterer,* $\sigma^{obj} = 0.001$ S/m, $SNR = 35$ dB) – Behaviour of the reconstruction error versus the relative permittivity of the object.

Towards this end, Figure 3 compares the reconstruction errors made by the *FH-MT-BCS* and the *FH-ST-BCS* methods when dealing with the reconstruction of two separated sparse targets whose relative permittivity is varied in the range $\varepsilon_r^{obj} \in [4.5, 7.0]$ [$\sigma^{obj} = 0.001$ S/m - Fig. 4(*a*)] and considering noisy data with $SNR = 35$ dB.

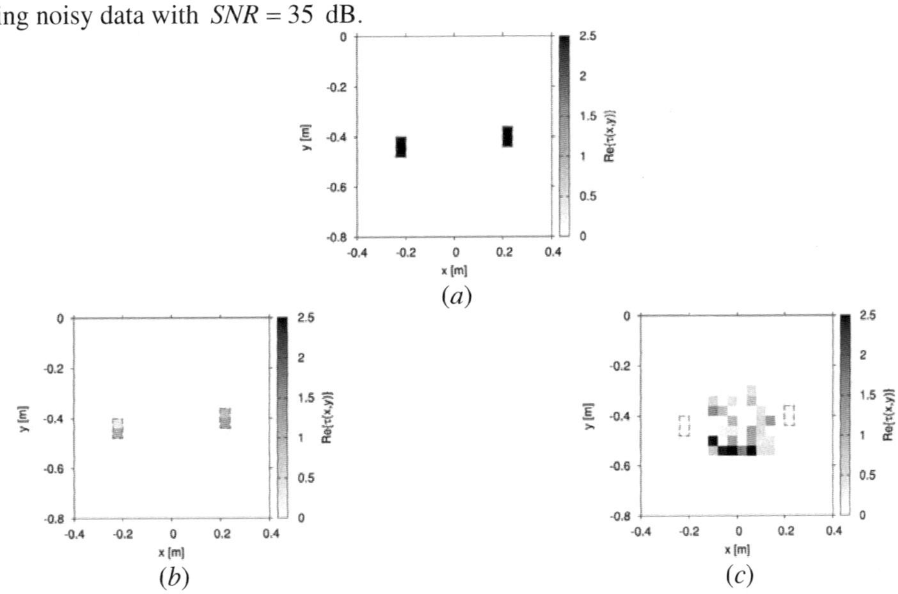

Figure 4. *Numerical Assessment* (*Two bars scatterer,* $\varepsilon_r^{obj} = 6.5$, $\sigma^{obj} = 0.001$ S/m, $SNR = 35$ dB) – (*a*) actual dielectric profile and retrieved solutions by the (*b*) *FH-MT-BCS* and (*c*) *FH-ST-BCS* techniques.

As it can be observed, lower errors are obtained by the *FH-MT-BCS* in all cases (Fig. 3). As a matter of fact, significantly better reconstructions are provided by the proposed method despite the non-negligible amount of noise, as it can be seen by looking at the retrieved profiles in Fig. 4 for $\varepsilon_r^{obj} = 6.5$. Even if providing a slight underestimation of the contrast, the *FH-MT-BCS* is able to correctly detect the scatterer support [Fig. 4(*b*)], while the single-task version yields an unsatisfactory result [Fig. 4(*c*)].

4. Conclusions

An innovative inverse scattering technique has been presented in order to deal with the retrieval of the dielectric characteristics of a buried investigation domain starting from wide-band *GPR* measurements. The proposed *FH-MT-BCS* methodology efficiently combines the information coming from multi-chromatic data thanks to an innovative initialization strategy of the *RVM* solver, as well as benefits from the regularization capabilities of the *MT-BCS* solver. The main novelty of the work is the innovative integration of the *MT-BCS* within the *FH* strategy in order to progressively exploit the acquired information about the solution at the low-pass reconstructions, thanks to the introduction of a customized "constrained" *RVM* solver. Some numerical results have been presented, in order to validate the effectiveness of the *FH-MT-BCS*, as well as to show its superior performances with respect to a single task-based solution within the same framework. Future work will be devoted at extending the developed algorithm to deal with more realistic scenarios by considering, for example, a non-static behaviour of both object and soil electromagnetic properties, the presence of metallic/PEC scatterers, as well as a full three-dimensional formulation of the subsurface scattering equations.

References

[1] Lesselier D and Habashy T 2000 Special issue on electromagnetic imaging and inversion on the Earth's subsurface *Inverse Probl.* **16**

[2] Chew W C and Lesselier D 2004 Special issue on electromagnetic characterization of buried obstacles *Inverse Probl.* **20**

[3] Chen C-C, Johnson J T, Sato M and Yarovoy A G 2007 Special issue on subsurface sensing using ground-penetrating radar *IEEE Trans. Geosci. Remote Sens.* **45** 2419–2421

[4] Peters L, Daniels J and Young J D 1994 Ground penetrating radar as a subsurface environmental sensing tool *IEEE Proc.* **82** 1802–1822

[5] Soldovieri F, Solimene R, Lo Monte L and Bavusi M 2011 Sparse reconstruction from GPR data with applications to rebar detection *IEEE Trans. Instr. Meas.* **60** 1070-1079

[6] Salucci M, Oliveri G and Massa A 2015 GPR prospecting through an inverse scattering frequency-hopping multifocusing approach *IEEE Trans. Geosci. Remote. Sens.* **53** 6573-6592

[7] Salucci M, Poli L and Massa A 2017 Advanced multi-frequency GPR data processing for non-linear deterministic imaging *Signal Processing* **132** 306-318

[8] Persico R and Sodovieri F 2008 Effects of background removal in linear inverse scattering *IEEE Trans. Geosci. Remote Sens.* **46** 1104-1114

[9] Rocca P, Benedetti M, Donelli M, Franceschini D and Massa A 2009 Evolutionary optimization as applied to inverse scattering problems *Inverse Probl.* **25** 1-41

[10] Salucci M, Poli L, Anselmi N and Massa A 2017 Multifrequency particle swarm optimization for enhanced multiresolution GPR microwave imaging *IEEE Trans. Geosci. Remote Sens.*, in press

[11] Massa A, Rocca P and Oliveri G Compressive sensing in electromagnetics - A review *IEEE Antennas Propag. Mag.* **57** 224-238

[12] Anselmi N, Oliveri G, Hannan M A, Salucci M and Massa A 2017 Color compressive sensing imaging of arbitrary-shaped scatterers *IEEE Trans. Microw. Theory Techn.*, in press

[13] Anselmi N, Oliveri G, Salucci M and Massa A 2015 Wavelet-based compressive imaging of sparse targets *IEEE Trans. Antennas Propag.* **63** 4889-4900

Acknowledgment

This work has been partially supported by the SIRENA project (2014-2017) funded by DIGITEO (France) under the "Call for Chairs 2014" and benefited from the networking activities carried out within the EU funded COST Action TU1208 "Civil Engineering Applications of Ground Penetrating Radar".

IOP Conf. Series: Journal of Physics: Conf. Series **904** (2017) 012003 doi:10.1088/1742-6596/904/1/012003

An inversion strategy for energy saving in smart building through wireless monitoring

N Anselmi[1] and T Moriyama[2]

[1] ELEDIA Research Center (ELEDIA@UniTN - University of Trento)
Via Sommarive 9, I-38123 Trento, Italy
[2] ELEDIA Research Center (ELEDIA@UniNAGA - University of Nagasaki)
852-8521, Nagasaki, Japan
E-mail: {nicola.anselmi, toshifumi.moriyama}@eledia.org

Abstract. The building plants represent one of the main sources of power consumption and of greenhouse gases emission in urban scenarios. The efficiency of energy management is also related to the indoor environmental conditions that reflect on the user comfort. The constant monitoring of comfort indicators enables the accurate management of building plants with the final objective of reducing energy waste and satisfying the user needs. This paper presents an inversion methodology based on support vector regression for the reconstruction and forecasting of the thermal comfort of users starting from the indoor environmental features of the building. The environmental monitoring is performed by means of a wireless sensor network, which pervasively measures the spatial variability of indoor conditions. The proposed system has been experimentally validated in a real test-site to assess the advantages and the limitations in supporting the management of the building plants towards energy saving.

1. Introduction

The need of more energy efficient and sustainable buildings has gained more and more attention in the last decades. Climate changes like the increasing global temperatures and the weather alterations are considered some of the consequences of the gas emissions originated by the combustion of fossil fuel. The energy generation from renewable resources and a more accurate management of consumptions are often imposed to limit the impact on the environment.

Private and public buildings are responsible of about one third of the urban energy consumptions [1][2]. Therefore, it is evident that the optimization of energy usage at building level would significantly impact on the power demand. Toward this end, besides the introduction of new materials and more efficient construction strategies, one of the main requirements is to know how the building systems behave and their effects on the satisfaction of the end-users.

Among the innovative solutions for the distributed monitoring of indoor spaces, the wireless sensor network (WSN) technology is suitable for the scalable and low-cost measurement of heterogeneous environmental parameters. The potentialities of this wireless technology have been verified in a huge number of indoor/outdoor applications.

State of the art examples include WSN-based architectures for smart home applications [3][4], and wireless sensors for the monitoring of appliance consumptions and schedule in residential apartments [5]. Similar WSN platforms have been also exploited for the localization and tracking of both passive [6]-[8] and active [9] targets. Additional information about the indoor occupancy level has been also estimated from the analysis of environment perturbations [10].

The building information that can be measured with WSN nodes become fundamental for the evaluation of the key performance indicators (KPI) related to the quality of the indoor environment and the consequent user comfort. The changes in the KPI behaviour are mainly due to the weather conditions and to the functioning of the heating-ventilation-air-conditioning (HVAC) systems [11][12]. The standard control strategies of HVAC systems are based on measurement of local temperature probes and aim to preserve a predefined value or preconfigured temperature masks. Such an approach does not consider predictive control schemes to address demand-side flexibility (e.g., short-term variations like weather changes or indoor occupancy).

In this paper, a learning-by-example (LBE) inversion strategy based on support vector regression (SVR) is aimed at reconstructing the desired thermal comfort KPI [the predicted mean vote (PMV) [13] has been considered as standard comfort indicator] from the environmental conditions. The method exploits the weather forecasts for the short-term prevision of outdoor changes, which impact on the indoor conditions. The output of the approach provides the target temperatures that satisfy the PMV requirements. Accordingly, the HVAC system can be calibrated in order to optimize the problem unknowns (i.e., the indoor environmental parameters).

The proposed solution has been validated in a real test field using environmental measurements acquired by a WSN prototype and by a weather station installed in proximity of the building. A selected set of preliminary results are presented to show the capability to learn and forecast the indoor thermal conditions subject to the KPI requirement.

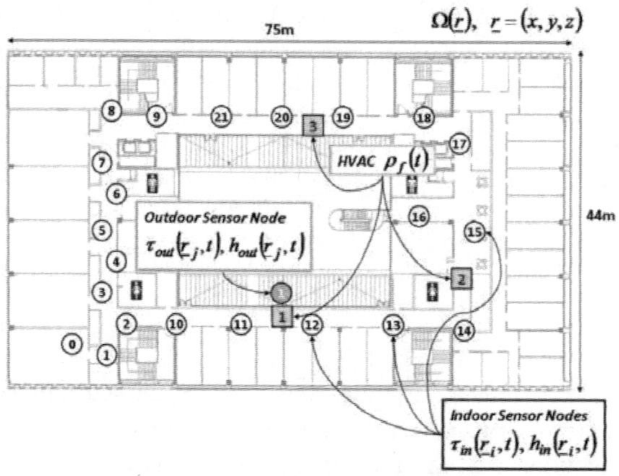

(*a*)

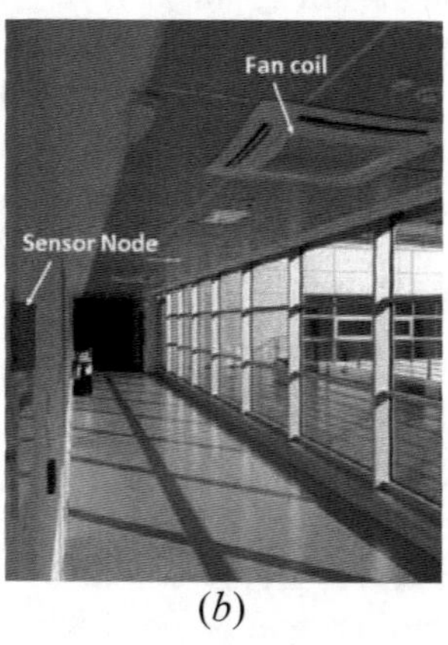

(*b*)

Figure 1. Investigation domain and system deployment: (*a*) floor map and node position, (*b*) real scenario.

2. System Architecture

Let us consider a WSN deployed in the three-dimensional domain $\Omega(\underline{r})$, $\underline{r}=(x,y,z)$ being the position vector. The WSN is composed by a set of $N=N_{in}+N_{out}$ nodes in positions $\underline{r}_i^{in}$, $i=1,...,N_{in}$, and $\underline{r}_j^{out}$, $j=1,...,N_{out}$, where N_{in} is the number of indoor nodes and N_{out} the number of nodes installed outdoor.

The outdoor nodes provide the reference environmental condition and together with the weather station allow the estimation of the complex 'indoor-outdoor transfer function' determined by the building properties. It has to be noticed that the weather station provides also the forecast of the outdoor temperature.

The WSN nodes are equipped with multiple sensors for the acquisition of the environmental parameters including the temperature $\tau_{in}(\underline{r}_i,t)$ and $\tau_{out}(\underline{r}_j,t)$, and the humidity $h_{in}(\underline{r}_i,t)$ and $h_{out}(\underline{r}_j,t)$, where t is the time instant of the data acquisition.

A finite set of F fan coil units are installed in known positions $\underline{r}_f$, $f=1,...,F$, as part of the HVAC system. The on/off working schedule of the fan coils is controlled by the main building management system and it is defined by the binary functions $\rho_f(t)=\{0,1\}$, $f=1,...,F$ [$\rho_f(t)=1$ if the *f*-th fan is active, $\rho_f(t)=0$, otherwise].

Starting from the indoor environmental parameters, the comfort indicators $PMV(\underline{r}_i,t)\in[-3;+3]$, $i=1,...,N_{in}$, are computed according to the formulation reported in [14]. It is worth noting that the PMV indicator includes additional parameters like the air velocity and the clothing insulation (more difficult to be measured and often initialized with reference values) to provide an evaluation of the comfort ranging from the cold [$PMV(\underline{r},t)=-3$] up to the hot [$PMV(\underline{r},t)=+3$] thermal sensation, passing through the ideal comfort value $PMV(\underline{r},t)=0$.

The proposed SVR-based learning technique is aimed at estimating the relation between the indoor and the outdoor environmental conditions in order to successively exploit weather forecast for the prevision of indoor conditions in the building positions $\underline{r}_i$, $i=1,...,N_{in}$. Toward this end, a set of regression functions $\Psi_{i,j}, i=1,...,N_{in}$, $j=1,...,N_{out}$ are trained for each *i-j* couple of nodes (a total of $N_{in}\times N_{out}$ functions are obtained) starting from the knowledge of the known samples (called training sets)

$$\left\{\left[\tau_{out}(\underline{r}_j,t),h_{out}(\underline{r}_j,t),PMV(\underline{r}_i,t),\rho_f(t),\tau_{in}(\underline{r}_i,t)\right]_{(i,j)};t=1,...,T\right\} \quad (1)$$

acquired at T different and representative time instants.

The function $\rho_f(t)$ is selected according to the rule $\min_f\left[\left|\underline{r}_f-\underline{r}_i\right|^2\right]$ in order to select the fan coil closer to the measurement node. After the training phase, the regression functions

$$\Psi_{i,j}:\left[\tau_{out}(\underline{r}_j,t),h_{out}(\underline{r}_j,t),PMV(\underline{r}_i,t),\rho_f(t)\right]\rightarrow\tau_{in}(\underline{r}_i,t) \quad (2)$$

are defined and successively exploited to estimate indoor temperatures at any point in time $t=t+\Delta t$, where Δt is the forecasting period. Moreover, it has to be noticed that the input feature $PMV(\underline{r}_i,t)$ can be used as requirement of the desired comfort in position $\underline{r}_i$.

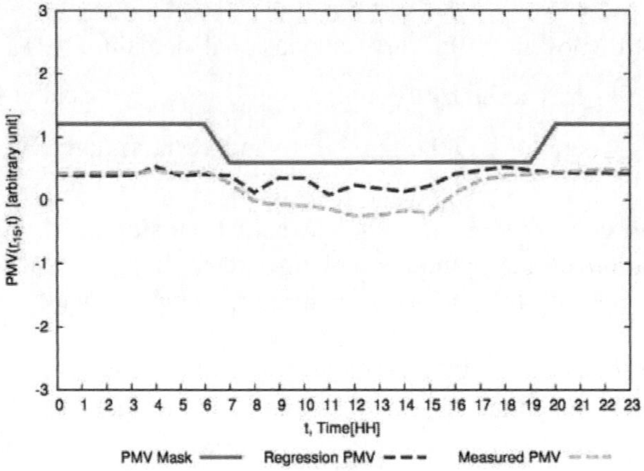

(*a*)

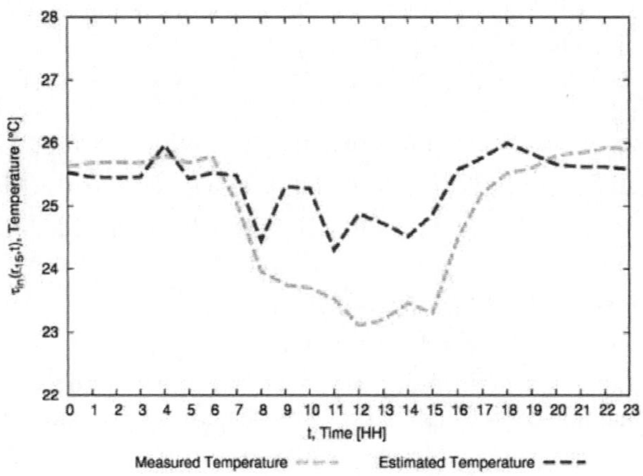

(*b*)

Figure 2. Summer period. (*a*) PMV mask vs measured and regression PMV. (*b*) measured vs estimated temperature.

3. Experimental Validation

The proposed system has been experimentally validated in a real test site at the ELEDIA Research Center of the University of Trento, Italy. More in detail, $N_{in} = 21$ sensor nodes distributed within $\Omega(\underline{r})$ at height $z_i = 2m$, $i = 1,...,N_{in}$ have been considered for indoor data acquisition. As regards the outdoor parameters, $N_{out} = 1$ node has been deployed. All the WSN nodes are equipped with SHT21 temperature and humidity sensors. Additional details about the wireless network architectures for smart home monitoring and management are reported in [3]. A long-term measurement campaign has been performed in order to collect environmental information in different seasons. Starting from such indoor features, the corresponding *PMV* indexes have been calculated.

Concerning the HVAC system, $F = 3$ representative fan coils have been monitored to extract the binary functions $\rho_f(t)$. For the sake of simplicity, the working schedule of the $f = 1,...,F$ fan coils is assumed equal.

The SVR methodology has been trained considering a set of $T = 216$ data samples acquired at different day and night hours, and both in Winter and in Summer periods. As a preliminary experiment, one regression function $\Psi_{15,1}$ related to the nodes $i=15$, $\underline{r}_{15}^{in}=\left(x_{15}^{in},y_{15}^{in},z_{15}^{in}\right)$ and $j=1$, $\underline{r}_{1}^{out}=\left(x_{1}^{out},y_{1}^{out},z_{1}^{out}\right)$ has been trained.

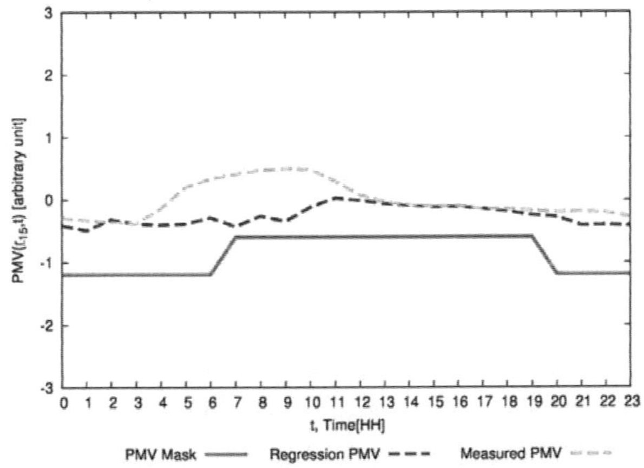

(*a*)

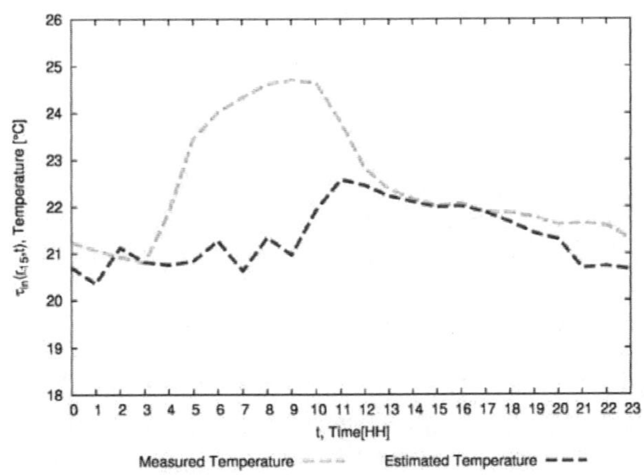

(*b*)

Figure 3. Winter period. (*a*) PMV mask vs measured and regression PMV. (*b*) measured vs estimated temperature.

During the test phase, the desired *PMV* daily mask has been defined and given as input feature of the test dataset. Different *PMV* masks have been designed for the summertime [Fig. 2(*a*)] and for the wintertime [Fig. 3(*a*)]. The objective of such comfort masks is to impose a daily comfort level that satisfies the user need in the desired time windows.

In order to validate the forecasting capability of the proposed approach, the outdoor parameters $\tau_{out}(\underline{r}_1,t)$ and $h_{out}(\underline{r}_1,t)$ with $t=t+\Delta t$, and $\Delta t=6h$, have been adopted exploiting the outdoor weather station. Such values have been used to create the test dataset. The obtained estimations of the predicted *PMV* are shown in Fig. 2(*a*) and Fig. 3(*a*) for the Summer and the Winter tests, respectively. The comparison with the measured *PMV* (i.e., the comfort obtained without the application of the proposed method) points out that the imposition of the comfort masks has caused higher (in Summer) and lower (in Winter) *PMV* values, closer to the desired and more energy-efficient values. Accordingly, the corresponding indoor temperatures are shown in Fig. 2(*b*) and Fig. 3(*b*).

IOP Conf. Series: Journal of Physics: Conf. Series **904** (2017) 012003 doi:10.1088/1742-6596/904/1/012003

4. Conclusions

In this paper, a novel methodology for the estimation and forecasting of thermal comfort in smart buildings has been proposed. The energy efficiency of HVAC systems has been addressed through the imposition of desired PMV masks. The corresponding indoor temperatures have been estimated as energy-efficient suggestion for the building management systems. The obtained results have shown the generalization capability of the SVR-based learning approach when dealing with real environmental measurements. Current activities are focused on the integration of location-based information about the user presence and behaviour within the monitored environments, in order to improve the management of HVAC plants according to the actual user needs. The wireless localization solutions presented in [3] will be exploited using the same wireless architectures adopted for environmental monitoring.

References

[1] Berge B 2009 Ecology of building materials *2nd Edn Elsevier Oxford.*

[2] US Congress Office of Technology Assessment 1991 Energy in developing countries- OTA-E-486 *U.S. Government Printing Office, Washington, DC.*

[3] Viani F, Robol F, Polo A, Rocca P, Oliveri G, and Massa A 2013 Wireless architectures for heterogeneous sensing in smart home applications – concepts and real implementations *IEEE Proc.* **101** 2381-2396

[4] Viani F, Polo A, Garofalo P, Anselmi N, Salucci M, and Giarola E 2017 Evolutionary optimization applied to wireless smart lighting in energy-efficient museums *IEEE Sensors Journal* **17** 1213-1214

[5] Viani F 2015 A decision support system based on wireless power metering for energy saving in smart buildings *Microw. Opt. Techn. Lett.* **57** 2750-2752

[6] Viani F, Lizzi L, Rocca P, Benedetti M, Donelli M, and Massa A 2008 Object tracking through RSSI measurements in wireless sensor networks *Electron. Lett.* **44** 653-654

[7] Viani F, Rocca P, Benedetti M, Oliveri G, and Massa A 2010 Electromagnetic passive localization and tracking of moving targets in a WSN-infrastructured environment *Inverse Problems* **26** 1-15

[8] Viani F, Rocca P, Oliveri G, Trinchero D, and Massa A 2011 Localization, tracking, and imaging of targets in wireless sensor networks: An invited review *Radio Science* **46**

[9] Ahmadi H, Polo A, Moriyama T, Salucci M, and Viani F 2016 Semantic wireless localization of WiFi terminals in smart buildings *Radio Science* **51** 876-892

[10] Viani F 2015 Opportunistic occupancy estimation in museums through wireless sensor networks *Microw. Opt. Techn. Lett.* **57** 1975-1977

[11] Ferreira P M, Silva S M, Ruano A E, Negrier A T, and Conceicao E Z E 2012 Neural network PMV estimation for model-based predictive control of HVAC systems *IEEE World Conf. on Computational Intelligence* (WCCI 2012), Brisbane, Australia, June 10-15

[12] Bhattacharya S, Sridevi S, and Pitchaiah R 2012 Indoor air quality monitoring using wireless sensor network *Int. Conf. on Sensing Technology* (ICST 2012), Kolkata, India, December 18-21

[13] Fanger P O 1972 Thermal comfort: analysis and applications in environ- mental engineering *New York: McGraw-Hill*

[14] Kang J, Kim Y, Kim H, Jeong J, and Park S 1997 Comfort sensing system for indoor environment *International Conference on Solid State Sensors and Actuators* **1** 311-314

Acknowledgment

This work has been partially supported by the Project "Zero Energy Buildings in Smart Urban Districts" (CTN01_00034_594053) within the National Technological Cluster on Smart Communities, funded by the Italian Ministry of Education, University, and Research (MIUR).

IOP Conf. Series: Journal of Physics: Conf. Series **904** (2017) 012004 doi:10.1088/1742-6596/904/1/012004

Indoor detection of passive targets recast as an inverse scattering problem

G Gottardi[1] and T Moriyama[2]

[1] ELEDIA Research Center (ELEDIA@UniTN - University of Trento)
Via Sommarive 9, I-38123 Trento, Italy
[2] ELEDIA Research Center (ELEDIA@UniNAGA - University of Nagasaki)
852-8521, Nagasaki, Japan
E-mail: {giorgio.gottardi, toshifumi.moriyama}@eledia.org

Abstract. The wireless local area networks represent an alternative to custom sensors and dedicated surveillance systems for target indoor detection. The availability of the channel state information has opened the exploitation of the spatial and frequency diversity given by the orthogonal frequency division multiplexing. Such a fine-grained information can be used to solve the detection problem as an inverse scattering problem. The goal of the detection is to reconstruct the properties of the investigation domain, namely to estimate if the domain is empty or occupied by targets, starting from the measurement of the electromagnetic perturbation of the wireless channel. An innovative inversion strategy exploiting both the frequency and the spatial diversity of the channel state information is proposed. The target-dependent features are identified combining the Kruskal-Wallis test and the principal component analysis. The experimental validation points out the detection performance of the proposed method when applied to an existing wireless link of a WiFi architecture deployed in a real indoor scenario. False detection rates lower than 2 [%] have been obtained.

1. Introduction

Many application fields related to security, smart home services [1], healthcare [2], benefit from the detection and localization of device-free targets. Several detection technologies have been investigated including wireless sensor networks (WSNs) [3]-[7], ultra-wideband (UWB) [8], radio frequency identification (RFID) [8], and wireless local area networks (WLANs) [10].

Recently, the channel state information (CSI) used by WLANs in the framework of orthogonal frequency division multiplexing (OFDM) has been exploited to enable location-based services. The CSI is a fine-grained feature of the channel frequency response (CFR) providing both magnitude and phase information at multiple frequencies [11]. This indicator is highly sensitive to the environmental changes and its information content can be exploited to characterize the target signature on the electromagnetic field. The multiple-input multiple output (MIMO) architecture has been introduced by the current WLAN standards to further improve the quality of the data transmission. The intrinsic spatial diversity of a MIMO system has been also exploited to enhance the performance of the wireless localization problem.

Many techniques exploiting the prominent features of the CSI exist in the state of the art, but the indoor scenario and the system setups are often dedicated to the localization itself. On the contrary, the proposed technique aims to introduce the detection capability on top of an existing wireless architecture already used for communication.

The goal of the proposed method is to provide a robust target detection only using a single pair of transmitter-receiver, which is the minimum set of hardware required to enable a wireless communication. The main contributions of the proposed solution include the definition of a target-dependent feature of the CSI, which is robust across the whole frequency spectrum, and the improvement of the detection performance respect to the state of the art (false detection rates lower than 2 [%] with a single pair of devices).

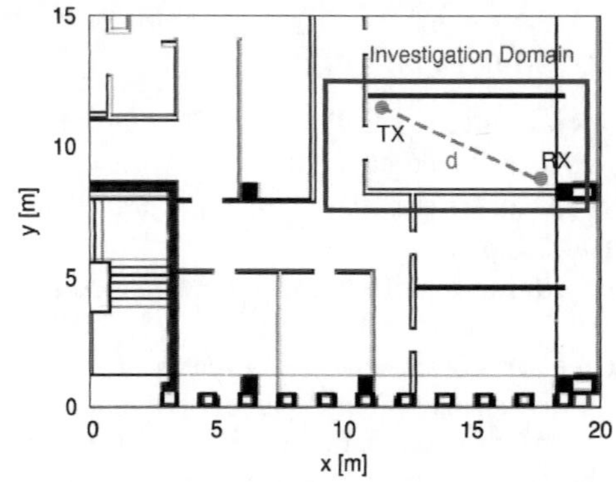

(*a*)

(*b*)

Figure 1. Investigation domain located in a standard office-area (*a*), and commodity WiFi receiver with *L=3* antennas (*b*).

2. Target Detection Strategy

The wireless channel measurement at multiple and orthogonal frequency sub-carriers is enabled by the OFDM technique adopted by the IEEE 802.11a/b/n wireless standard. The channel properties are represented in the form of the CSI. More in detail, the received signal can be defined as [12]

$$\underline{Q}(t)=\underline{\underline{H}}(t)\underline{P}(t)+\underline{\beta}(t) \tag{1}$$

where $\underline{Q}(t)=[q_k(t);k=1,...,K]$ and $\underline{P}(t)=[p_k(t);k=1,...,K]$ are the time-domain representation of the received and the transmitted complex-valued signals, respectively, K is the number of frequency sub-carriers, $\underline{\beta}(t)=[\beta_k(t);k=1,...,K]$ is the noise vector, and t is the measurement time. Moreover, $\underline{\underline{H}}(t)$ is the multi-frequency channel matrix represented as

IOP Conf. Series: Journal of Physics: Conf. Series **904** (2017) 012004 doi:10.1088/1742-6596/904/1/012004

$$\underline{\underline{H}}(t)=\left[\underline{\underline{H}}_1(t),\underline{\underline{H}}_2(t),\ldots,\underline{\underline{H}}_K(t)\right]^T \tag{2}$$

where $\underline{\underline{H}}_k(t)$ is a complex matrix with elements $h_{il}^{(k)}(t)=\left|h_{il}^{(k)}(t)\right|e^{j\angle h_{il}^{(k)}(t)}$, $i=1,\ldots,I$ and $l=1,\ldots,L$ being the transmitting and the receiving antennas, respectively, and T is the transpose symbol. The magnitude $\left|h_{il}^{(k)}(t)\right|$ $[dB]$ and the phase $\angle h_{il}^{(k)}(t)$ $[deg]$ are inferred from the knowledge of the transmitted and the received signals when the i-th antenna is transmitting and the l-th antenna is receiving. The fine-grained channel frequency response given by the matrix $\underline{\underline{H}}(t)$ includes both frequency and spatial diversity information, since $W = I \times L$ wireless links exist between the transmitter and the receiver.

Only the CSI amplitude has been exploited for detection purposes. The phase information is fundamental but the calibration procedure has been avoided in this work to reduce as much as possible the computational complexity of the method. The processing of the phase is under study to make the such information stable and usable for detection purposes. Up to now, the CSI phase has introduced a very high sensitivity that is less compatible with the robustness objective addressed in the proposed detection method. The attention has been focused on the exploitation of the spatial and frequency diversity to maximize the robustness of the target detection even in complex and real-world scenarios, as opposite to several solutions in the state of the art that apply data fusion strategies to combine the multiple links and carriers in a single aggregated data stream.

In order to solve the inverse problem, the time evolution of the CSI has been analyzed to recognize the presence of patterns in the received signal due to the target presence in the domain. The probability density function (PDF) of the CSI correlations $C_{k,w}^{(void)}$ (i.e., the correlation between two CSI streams acquired without targets) and $C_{k,w}^{(full)}$ (i.e., the correlation between one CSI stream without targets and one CSI stream with targets) has been considered for all the carriers $k=1,\ldots,K$ and the links $w=1,\ldots,W$ to detect the pattern changes in the data acquired in absence and in presence of the targets. The *absence* data set has been acquired when the detection domain is empty, while the *presence* data set has been collected when targets are within the area of interest. The comparison between the PDFs of the correlations has been performed using the non-parametric Kruskal-Wallis (KW) test to statistically quantify the similarity of the distributions. The KW test has been adopted since the strong assumption that the PDFs of the correlations are normally distributed is not required. The p-values $\rho_{k,w}$, $k=1,\ldots,K$, $w=1,\ldots,W$ returned by the KW test is a measure of statistical similarity. High p-values stand for distributions of the same population, whereas lower p-values are obtained when at least one sample of the distribution is significantly different from the others. The computed p-values are the set of features sensitive to the target presence, and rapidly decrease in presence of targets.

2.1. *Target-dependent features selection*

The positions of the multiple transmitting and receiving antennas of the MIMO link determine different propagation paths and consequently the mutual correlation among the W wireless links can be leveraged for the solution of the detection problem. The perturbations due to the presence of a human body introduce noisy but highly correlated changes of amplitudes across the W wireless links. The main challenge is to identify the components of the perturbations that are target-dependent. Toward this end, the PCA method [12] has been adopted to select the orthogonal components of the p-values not belonging to the noise subspace. The computed principal components $\underline{\gamma}_{\hat{k}}=\left[\gamma_{\hat{k},w};w=1,\ldots,W\right]=\Omega\left[\rho_{\hat{k},w};w=1,\ldots,W\right]$, $\hat{k}=1,\ldots,K$, where $\Omega[\cdot]$ computes the PCA of the p-values, are ordered so that the first ones account for most of the data variability. The goal is to identify the PCA components more related to the target absence/presence. Accordingly, it has been verified in

IOP Conf. Series: Journal of Physics: Conf. Series **904** (2017) 012004 doi:10.1088/1742-6596/904/1/012004

[13] that the first components (i.e., the ones with the higher eigenvalues) can be discarded in order to remove the intrinsic noise of the input data and to isolate the smaller variations.

The binary status of target absence/presence has been inferred applying the following thresholding strategy

$$\chi_w = \begin{cases} 1 \ \ if \ \ \frac{1}{K}\sum_{k=1}^{K}\gamma_{k,w} < \gamma_{th} \\ 0 \ \ \ \ \ \ \ \ \ \ otherwise \end{cases} ; \ w = 1,...,W \tag{3}$$

to the set of principal components of each wireless link. The components more related to the actual target absence/presence are selected as the target-dependent features.

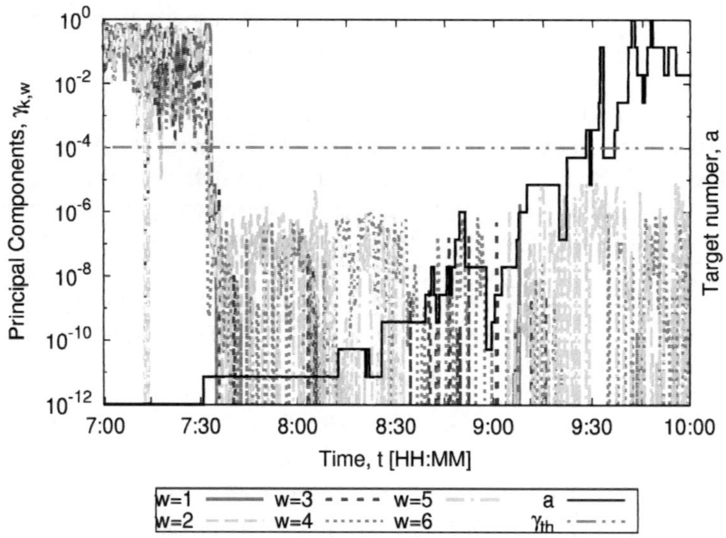

Figure 2. Temporal trend of the principal components of the CSI computed in absence and in presence of targets in the investigation domain.

3. Experimental Validation

The experiments have been carried out using commodity WiFi devices deployed in indoor areas. The WiFi access point TP-Link AC1750 compliant to the IEEE 802.11n standard has been adopted as transmitter operating at frequency $f = 2.4$ [GHz], and the Intel 5300 NIC has been used for the acquisition of the CSI at the receiver [Fig. 1(*b*)].

Multiple wireless links ($W = 6$) have been considered, since $I = 2$ transmitting and $L = 3$ receiving antennas have been assumed. The sampling rate of the receiver is $t = 0.5$ [s], which is the default rate of the ping command adopted to detect network devices, so that the CSI acquisition is transparent in terms of network overhead). The data sets have been collected from 2:00 AM to 5:00 AM for the *absence* status, and during the daytime for the periods of target *presence*. The actual number of targets $a = 0, ..., A$ is inferred from the recordings taken by a video surveillance system installed ad-hoc to acquire the ground truth during the experiments.

The false positive rate $\omega_w^{(FP)}$ and the false negative rate $\omega_w^{(FN)}$ have been computed for each wireless link $w = 1,...,W$ starting from the comparison of the estimated target absence/presence in (3) and the actual conditions of the scenario at hand. Such performance indicators are representative of the detection robustness, defined as the stability and reliability of the estimations respect to the actual target absence and presence.

The detection has been performed in a classroom of size 40 [m^2] shown in Fig. 1(*a*). The wireless link between the transmitter and the receiver graphically represented in the map has a length of $d = 8$ [m].

The CSI data set is composed by an *absence* period from 7:00 AM to 7:30 AM, and successively by a *presence* period when one target entered the area until 8:10 AM. After that, a variable number of targets up to $A = 14$ occupied the classroom until 10:00 AM.

For the sake of conciseness, the principal components have been computed choosing the center subcarrier $k = 15$. Figure 2 shows the temporal trends of the obtained principal components $\gamma_{\hat{k},w}; w = 1,...,W$ of each wireless link $w=1,...,W$, $W=6$, together with the number of targets $a = 0, ...,A$, $A = 14$. The trends point out a rapid decrease of all the principal components when the first target occupies the area at 7:30 AM. The decrease is the result of an evident change between the PDFs regardless the wireless links $w = 1, ..., W$.

The false positive and false negative rates have been averaged over the wireless links pointing out detection performance of $\omega_w^{(FP)} = 1.17$ [%] and $\omega_w^{(FP)} = 0.16$ [%] with the threshold $\gamma_{th} = 10^{-4}$. Such a threshold has been calibrated computing the lower boundary $\bar{\gamma} - 3\sigma$, where $\bar{\gamma}$ is the average value of the principal components measured during the experimental acquisition in absence of targets, and σ is the variance. The obtained performance outperform other state of the art methods, like PADS [14] and FIMD [15], which show higher false rates (ROC curves point out false positive rates higher than 5 [%] with detection rates of 80 [%]).

4. Conclusions

In this paper, the robust detection of device-free targets using commodity WiFi devices has been addressed. The detection problem has been recast as an inverse problem since the properties of the investigation domain (i.e., the absence/presence of at least one target) have been reconstructed starting from the electromagnetic perturbation measured by one wireless link. The CSI data have been processed exploiting both the frequency and the spatial diversity of the MIMO wireless architecture and the proposed PCA-based method pointed out reliable detection performance with failure rates lower than 2 [%].

References

[1] Viani F, Polo A, Garofalo P, Anselmi N, Salucci M, and Giarola E 2017 Evolutionary optimization applied to wireless smart lighting in energy-efficient museums *IEEE Sensors Journal* **17** 1213-1214

[2] Viani F, Robol F, Polo A, Rocca P, Oliveri G, and Massa A 2013 Wireless architectures for heterogeneous sensing in smart home applications – concepts and real implementations *IEEE Proc. IEEE* **101** 2381-2396

[3] Patwari N and Wilson J 2010 RF sensor networks for device-free localization: Measurements, models, and algorithms *Proc. IEEE* **98** 1961-1973

[4] Viani F, Lizzi L, Rocca P, Benedetti M, Donelli M, and Massa A 2008 Object tracking through RSSI measurements in wireless sensor networks *Electron. Lett.* **44** 653-654

[5] Viani F, Rocca P, Benedetti M, Oliveri G, and Massa A 2010 Electromagnetic passive localization and tracking of moving targets in a WSN-infrastructured environment *Inverse Problems* **26** 1-15

[6] Viani F, Rocca P, Oliveri G, Trinchero D, and Massa A 2011 Localization, tracking, and imaging of targets in wireless sensor networks: An invited review *Radio Science* **46**

[7] Viani F 2015 Opportunistic occupancy estimation in museums through wireless sensor networks *Microw. Opt. Techn. Lett.* **57** 1975-1977

[8] McCracken M and Patwari N 2014 Indoor localization with range-based measurements and little prior information *IEEE Trans. Mobile Comput.* **13** 1509-1521

[9] Ni L M, Zhang D, and Souryal M R 2011 RFID-based localization and tracking technologies *IEEE Wireless Comm.* **18** 45-51

[10] Ahmadi H, Polo A, Moriyama T, Salucci M, and Viani F 2016 Semantic wireless localization of WiFi terminals in smart buildings *Radio Science* **51** 876-892

[11] Chapre Y, Ignjatovic A, Seneviratne A, and Jha S 2015 CSI-MIMO: an efficient Wi-Fi fingerprinting using channel state information with MIMO *Pervasive Mob. Comput.* **23** 89-103

[12] Wu K, Xiao J, Yi Y, Chen D, Luo X, Ni L 2013 CSI-based indoor localization *IEEE Trans. Parallel Distrib. Syst.* **24** 1300-1309

[13] Jolliffe I T 2002 Principal Component Analysis *Springer-Verlag, New York, 2nd Ed.*

[14] Qian K, Wu C, Yang Z, Liu Z, Zhou Z 2014 PADS: Passive detection of moving targets with dynamic speed using PHY layer information *20th IEEE Int. Conf. on Parallel and Distributed Systems* (ICPADS 2014), Hsinchu, Taiwan, 16-19 December.

[15] Xiao J, Wu K, Yi Y, Wang L, Ni L M 2012 FIMD: Fine-grained device-free motion detection *18th IEEE Int. Conf. on Parallel and Distributed Systems* (ICPADS 2012), Nanyang, Singapore, 17-19 December

IOP Conf. Series: Journal of Physics: Conf. Series **904** (2017) 012005 doi:10.1088/1742-6596/904/1/012005

Filtering techniques for efficient inversion of two-dimensional Nuclear Magnetic Resonance data

V. Bortolotti[1], L. Brizi[2,3], P. Fantazzini[2,3], G. Landi[4], F. Zama[4]

E-mail: fabiana.zama@unibo.it
[1]Department of Civil, Chemical, Environmental and Materials Engineering, University of Bologna, Italy.
[2]Department of Physics and Astronomy, University of Bologna, Italy.
[3]Centro Fermi e Museo Storico della Fisica e Centro Studi e Ricerche Enrico Fermi, Roma, Italy.
[4]Department of Mathematics, University of Bologna, Italy.

Abstract.
The inversion of two-dimensional Nuclear Magnetic Resonance (NMR) data requires the solution of a first kind Fredholm integral equation with a two-dimensional tensor product kernel and lower bound constraints. For the solution of this ill-posed inverse problem, the recently presented 2DUPEN algorithm [V. Bortolotti et al., *Inverse Problems*, 33(1), 2016] uses multi-parameter Tikhonov regularization with automatic choice of the regularization parameters. In this work, I2DUPEN, an improved version of 2DUPEN that implements Mean Windowing and Singular Value Decomposition filters, is deeply tested. The reconstruction problem with filtered data is formulated as a compressed weighted least squares problem with multi-parameter Tikhonov regularization. Results on synthetic and real 2D NMR data are presented with the main purpose to deeper analyze the separate and combined effects of these filtering techniques on the reconstructed 2D distribution.

1. Introduction

Nuclear Magnetic Resonance (NMR) of ^{1}H nuclei of hydrogenous fluids confined in porous media provides a set of techniques exploited to study pore space structure and fluids behavior in a wide range of systems from sedimentary rocks to biological materials [1]. Parameters like the longitudinal (T_1) and the transverse (T_2) relaxation times, as well as the self-diffusion molecular coefficient, can be determined and associated to properties of the fluids and porous media. When a hydrogenous fluid saturates a porous medium, T_1 and T_2 follow distributions of values and the problem is to correlate these two-dimensional distributions of relaxation times with the two-dimensional experimental data. The resulting two dimensional distributions are shown as two-dimensional relaxation maps characterized by peaks of different size and shape over flat regions. The problem of determining the realxation times distribution from NMR data is a well-known ill-posed inverse problem. In [2], the 2DUPEN algorithm has been presented, employing multi-parameter Tikhonov regularization and lower bound constraints, for the inversion of two-dimensional NMR relaxation data. In order to compute local values of the regularization parameter, 2DUPEN uses the *Uniform PENalty (UPEN) principle* [3] stating that the product of the regularization parameter and the curvature value in each point of the relaxation times

IOP Conf. Series: Journal of Physics: Conf. Series **904** (2017) 012005 doi:10.1088/1742-6596/904/1/012005

distribution should be constant. Therefore, 2DUPEN is an iterative algorithm where, at each iteration, the locally adapted regularization parameters are automatically updated using the UPEN principle, and an approximate distribution is obtained by solving a L_2-regularized least squares problem subject to lower bound constraints, usually non negativity constraints. The Newton Projection method [4] is used for the solution of the constrained subproblem. The results of numerical experiments on synthetic and real data show that 2DUPEN is effective in reconstructing peaks and flat regions of the unknown distribution with the same accuracy. However, when dealing with real 2D NMR relaxation data, 2DUPEN may require a huge computational work since it involves solving a sequence of large scale non-negatively constrained least squares problems. Therefore it is necessary to reduce the computation effort both from the point of view of the storage requirement and the computation time. In this work, this aim is obtained by introducing and combining two classical techniques, namely the Singular Value Decomposition (SVD) and the Mean Windowing (MW) methods. The application of SVD for data compression has a long history in inverse problems and also in NMR data inversion. In [5] Sezginer used the SVD method to compress the echo train data and kernel matrix. In [6] echo train data and kernel matrix compression are extended to inverse problems with tensor product structure. In that work a preliminary step applies SVD to reduce problem size by projecting the data and kernels onto a lower dimensional subspace. The main difference between [6] and our method is the employing of the uniform penalty principle to compute the locally adapted regularization parameters whereas in [6] a scalar global regularization parameter is used. Hence two different constrained least squares models are solved. Moreover the projected Newton method is used in the present work to solve the constrained minimization problem where in [6] the BRD [7] method is used. The MW method has been used in [3] to compress the data blocks considerably reducing the amount of data. In [8], a preliminary use of MW and SVD filters have been successfully tested to improve the computational cost of 2DUPEN when applied to *real* 2D NMR data. This enhanced version of 2DUPEN was called I2DUPEN. The MW and SVD filters turned out to be effective in reducing the problem size and the computational effort of 2DUPEN with real data, thus motivating a deeper analysis of these filters on simulated data. Hence, the main purpose of this paper is to analyse the separate and combined effects of these filtering techniques on synthetic and real 2D NMR data. Our experimental analysis shows that the combined use of the MW and SVD filters has a considerable effect on reducing the computational cost of 2DUPEN while preserving the reconstruction quality. In particular I2DUPEN preserves the principal feature of UPEN algorithm, that is to reproduce the underlying distribution in presence of peaks and flat regions. The sequel is organized as follows: in Section 2 the problem of the inversion of 2D NMR data is described. Section 3 presents the filtering techniques used in the 2DUPEN algorithm and numerical results are reported in Section 4. Finally, conclusions are given in Section 5.

2. Problem description

If we consider 2D NMR relaxation data acquired using a conventional Inversion-Recovery (IR) or Saturation-Recovery (SR) experiment detected by a Carr-Purcell-Meiboom-Gill (CPMG) pulse train [9], then the evolution time t_1 in IR or SR and the evolution time t_2 in CPMG are two independent variables and the 2D NMR relaxation data $S(t_1, t_2)$ can be expressed as:

$$S(t_1,t_2) = \iint_0^\infty k_1(t_1,T_1)k_2(t_2,T_2)F(T_1,T_2)\, dT_1\, dT_2 + e(t_1,t_2) \tag{1}$$

where the unknown $F(T_1, T_2)$ is the distribution of T_1 and T_2 relaxation times, the kernels have the following expressions:

$$k_1(t_1,T_1) = 1 - \gamma \exp(-t_1/T_1), \quad \gamma = \begin{cases} 1, & SR \\ 2, & IR \end{cases}, \quad k_2(t_2,T_2) = \exp(-t_2/T_2)$$

IOP Conf. Series: Journal of Physics: Conf. Series **904** (2017) 012005 doi:10.1088/1742-6596/904/1/012005

and $e(t_1, t_2)$ represents Gaussian additive noise. The unknown distribution $F(T_1, T_2)$ is known to be greater or equal to a constant $\rho \in \mathbb{R}$; in this work, $\rho = 0$. After considering $M_1 \times M_2$ samples of the times t_1 and t_2, and $N_1 \times N_2$ samples of the relaxation times T_1 and T_2, problem (1) can be discretized as:

$$(\mathbf{K}_2 \otimes \mathbf{K}_1)\mathbf{f} + \mathbf{e} = \mathbf{s} \tag{2}$$

where $\mathbf{K}_1 \in \mathbb{R}^{M_1 \times N_1}$, $\mathbf{K}_2 \in \mathbb{R}^{M_2 \times N_2}$ are the discretized exponential kernels, $\mathbf{s} \in \mathbb{R}^M$, $M = M_1 \cdot M_2$, is the discrete vector of the measured noisy data, $\mathbf{f} \in \mathbb{R}^N$, $N = N_1 \cdot N_2$, is the vector reordering of the distribution and $\mathbf{e} \in \mathbb{R}^M$ is the vector with the discretized noise. In order to avoid artifact peaks in the computed distribution, 2DUPEN uses multiple-parameter Tikhonov regularization and solves the minimization problem

$$\min_{\mathbf{f} \geq 0} \left\{ \|(\mathbf{K}_2 \otimes \mathbf{K}_1)\mathbf{f} - \mathbf{s}\|^2 + \sum_{i=1}^{N} \lambda_i (\mathbf{L}\mathbf{f})_i^2 \right\} \tag{3}$$

where $\| \cdot \|$ is the L_2 norm, $\mathbf{L} \in \mathbb{R}^{N \times N}$ is the discrete Laplacian operator and λ_i are the regularization parameters, $i = 1, \ldots, N$. The 2DUPEN method is an iterative procedure where, at each iteration, suitable values for the λ_i's are determined by imposing that all the non-null products $\lambda_i(\mathbf{L}\mathbf{f})_i^2$ have the same constant value (UPEN principle [2, 3]) and an approximated distribution is obtained by solving a problem of the form (3) with the Newton Projection method. 2DUPEN is able to achieve high quality reconstructions [2] but our experience indicates that considerable computational effort is needed. The computational effort is given by the cost of multiplications between the matrix $\mathbf{K}_2 \otimes \mathbf{K}_1$ (and its transpose) and vectors in the Newton Projection method (see [2] for details about the algorithm). Using the properties of the Kronecker products, the matrices $\mathbf{K}_2 \otimes \mathbf{K}_1$ and $(\mathbf{K}_2 \otimes \mathbf{K}_1)^T$ are never formed and the matrix vector products are computed as:

$$(\mathbf{K}_2 \otimes \mathbf{K}_1)\mathbf{x} = vec(\mathbf{K}_1 \mathbf{X} \mathbf{K}_2^T)$$

where $vec(\mathbf{V})$ is the vector obtained by columnwise reordering the matrix $\mathbf{V}$; therefore, in our case, $\mathbf{x} = vec(\mathbf{X})$ and $\mathbf{X} \in \mathbb{R}^{N_1 \times N_2}$. The matrix vector multiplications involving $(\mathbf{K}_2 \otimes \mathbf{K}_1)^T$ are performed using the relation $(\mathbf{K}_2 \otimes \mathbf{K}_1)^T = \mathbf{K}_2^T \otimes \mathbf{K}_1^T$. The storage requirement is therefore $(M_1 + M_2)N_1 + N_1^2$ bytes (assuming $N_1 = N_2$). Therefore an effective reduction of computation cost and storage requirements is obtained, in this work, by means of techniques that allow us to decrease the problems size while preserving properties of the Kronecker product matrices. For this reason we do not propose here methods, such as parallel block iterative algorithms, which require the formation of the matrix $\mathbf{K}_2 \otimes \mathbf{K}_1$, with storage requirement of $M_1 \cdot M_2 \cdot N_1 \cdot N_2$ bytes. Moreover the formation of $\mathbf{K}_2 \otimes \mathbf{K}_1$ would require the parallelization of matrix vector products involving major changes in the software and hardware devices usually available in real NMR data analysis applications.

3. Filtering techniques

In order to accelerate the implementation of the 2DUPEN method we explore projection methods (SVD filtering) and data reduction methods (MW technique) sparately and then we couple them together. In the following paragraphs we describe the different techniques and deeply analyze the Improved 2DUPEN algorithm (I2DUPEN) which includes both SVD and MW filters.

The Mean Windowing-like filter. The CPMG data blocks of an IR-CPMG or SR-CPMG sequence, used to acquire the two dimensional $T_1 - T_2$ experimental data, usually have thousands of points; in order to reduce their number, the CPMG data blocks are averaged on windows of logarithmically increasing size, decreasing the number of points by orders of magnitude. In order to take into account the number of points averaged in each window [3], weighting factors are introduced in the data fitting term of (3). Since each IR or SR point has the same weight, the weighting matrix can be expressed as a Kronecker product of diagonal matrices, i.e. $(\mathbf{B} \otimes \mathbf{I})$ where $\mathbf{B}$ contains the numbers of points of each CPMG windowed block, and $\mathbf{I}$ is the identity matrix. Assuming that the size of the whole 2D data is $M_1 \times M_2$, after the application of the windowing, the vector reordered data is $\overline{\mathbf{s}} \in \mathbb{R}^{M_1 \cdot \overline{M}_2}$ with $\overline{M}_2 \ll M_2$. Introducing the weighting matrix $(\mathbf{B} \otimes \mathbf{I})$, we obtain the following weighted least squares problem:

$$\min_{\mathbf{f} \geq 0} \left\{ \left\| (\overline{\mathbf{K}}_2 \otimes \mathbf{K}_1)\mathbf{f} - \overline{\mathbf{s}} \right\|^2_{(\mathbf{B} \otimes \mathbf{I})^2} + \sum_{i=1}^{N} \lambda_i (\mathbf{L}\mathbf{f})_i^2 \right\} \tag{4}$$

where $\overline{\mathbf{K}}_2 \in \mathbb{R}^{\overline{M}_2 \times N_2}$ is the t_2 kernel of the reduced size problem, $\mathbf{B} \in \mathbb{R}^{\overline{M}_2 \times \overline{M}_2}$. We have:

$$\left\| (\overline{\mathbf{K}}_2 \otimes \mathbf{K}_1)\mathbf{f} - \overline{\mathbf{s}} \right\|^2_{(\mathbf{B} \otimes \mathbf{I})^2} = \left\| (\mathbf{B} \otimes \mathbf{I}) \left[(\overline{\mathbf{K}}_2 \otimes \mathbf{K}_1)\mathbf{f} - \overline{\mathbf{s}} \right] \right\|^2. \tag{5}$$

Using (5) and the Kronecker product property $(\mathbf{B} \otimes \mathbf{I})(\overline{\mathbf{K}}_2 \otimes \mathbf{K}_1) = (\mathbf{B}\overline{\mathbf{K}}_2 \otimes \mathbf{K}_1)$, problem (4) can be written as:

$$\min_{\mathbf{f} \geq 0} \left\{ \left\| (\widetilde{\mathbf{K}}_2 \otimes \mathbf{K}_1)\mathbf{f} - \widetilde{\mathbf{s}} \right\|^2 + \sum_{i=1}^{N} \lambda_i (\mathbf{L}\mathbf{f})_i^2 \right\} \tag{6}$$

where $\widetilde{\mathbf{K}}_2 = \mathbf{B}\overline{\mathbf{K}}_2$ and $\widetilde{\mathbf{s}} = (\mathbf{B} \otimes \mathbf{I})\overline{\mathbf{s}}$.

The SVD filter. While the MW filter reduces the data size by replacing the points in a window with their average value, with the SVD filter it is possible to reduce the data dimensions by projecting $\mathbf{s}$ onto a lower-dimensional subspace spanned by the first left singular vectors of $\mathbf{K}_1$ and $\mathbf{K}_2$. Let $\mathbf{K}_1 = \mathbf{U}_1 \mathbf{\Sigma}_1 \mathbf{V}_1^T$ and $\mathbf{K}_2 = \mathbf{U}_2 \mathbf{\Sigma}_2 \mathbf{V}_2^T$ be the SVD of $\mathbf{K}_1$ and $\mathbf{K}_2$, where

$$\mathbf{U}_1 \in \mathbb{R}^{M_1 \times M_1},\ \mathbf{V}_1 \in \mathbb{R}^{N_1 \times N_1},\ \mathbf{\Sigma}_1 \in \mathbb{R}^{M_1 \times N_1},\quad \mathbf{U}_2 \in \mathbb{R}^{M_2 \times M_2},\ \mathbf{V}_2 \in \mathbb{R}^{N_2 \times N_2},\ \mathbf{\Sigma}_2 \in \mathbb{R}^{M_2 \times N_2}$$

with $\mathbf{U}_i, \mathbf{V}_i$ orthogonal matrices and $\mathbf{\Sigma}_i$ diagonal matrices of the singular values, $i = 1, 2$. Hence, the data fitting term of (3) can be written as:

$$\|(\mathbf{K}_2 \otimes \mathbf{K}_1)\mathbf{f} - \mathbf{s}\|^2 = \|(\mathbf{\Sigma}_2 \mathbf{V}_2^T \otimes \mathbf{\Sigma}_1 \mathbf{V}_1^T)\mathbf{f} - (\mathbf{U}_2^T \otimes \mathbf{U}_1^T)\mathbf{s}\|^2. \tag{7}$$

We use the Truncated SVD (TSVD) and consider only the SVD components corresponding to the largest singular values. Let τ be a positive parameter and let ρ_1, ρ_2 be the number of singular values of $\mathbf{\Sigma}_1, \mathbf{\Sigma}_2$ greater or equal to τ, we replace problem (3) with

$$\min_{\mathbf{f} \geq 0} \left\{ \|(\hat{\mathbf{K}}_1 \otimes \hat{\mathbf{K}}_2)\mathbf{f} - \hat{\mathbf{s}}\|^2 + \sum_{i=1}^{N} \lambda_i (\mathbf{L}\mathbf{f})_i^2 \right\} \tag{8}$$

where: $\hat{\mathbf{K}}_1 = \hat{\mathbf{\Sigma}}_1 \hat{\mathbf{V}}_1^T$, $\hat{\mathbf{K}}_2 = \hat{\mathbf{\Sigma}}_2 \hat{\mathbf{V}}_2^T$, $\hat{\mathbf{s}} = (\hat{\mathbf{U}}_2^T \otimes \hat{\mathbf{U}}_1^T)\mathbf{s}$ and

$$\hat{\mathbf{U}}_1 \in \mathbb{R}^{M_1 \times \rho_1}, \hat{\mathbf{V}}_1 \in \mathbb{R}^{N_1 \times \rho_1},\ \hat{\mathbf{\Sigma}}_1 \in \mathbb{R}^{\rho_1 \times \rho_1},\quad \hat{\mathbf{U}}_2 \in \mathbb{R}^{M_2 \times \rho_2}, \hat{\mathbf{V}}_2 \in \mathbb{R}^{N_2 \times \rho_2},\ \hat{\mathbf{\Sigma}}_2 \in \mathbb{R}^{\rho_2 \times \rho_2}.$$

Filtered 2DUPEN methods. The MW and SVD filters can be used separately or together obtaining the following modifications of the 2DUPEN algorithm. If the SVD filter is applied to problem (3) without MW data reduction, then we obtain a modified 2DUPEN method, named 2DUPENm_1, which applies 2DUPEN to solve problem (8). If the SVD filter is applied, the data are reduced by MW but the weights are not used in the data fitting term, we have a modified 2DUPEN method, named 2DUPENm_2, which uses 2DUPEN to solve a SVD filtered problem of the form (8) where $\hat{\mathbf{K}}_2$ is obtained by TSVD of the matrix $\overline{\mathbf{K}}_2$ and $\hat{\mathbf{s}} = (\hat{\mathbf{U}}_2^T \otimes \hat{\mathbf{U}}_1^T)\overline{\mathbf{s}}$. If the data are reduced by the MW filter, the weights are used in the data fitting term and no SVD filter is used, we obtain a modified 2DUPEN method, named 2DUPENm_3, which applies 2DUPEN to solve (6). Finally, the so-called Improved 2DUPEN (I2DUPEN) algorithm uses both the MW and SVD filters and solves a problem of the form (8) with weighting factors, where $\hat{\mathbf{K}}_2$ is obtained by TSVD of the matrix $\widetilde{\mathbf{K}}_2$ and $\hat{\mathbf{s}} = (\hat{\mathbf{U}}_2^T \otimes \hat{\mathbf{U}}_1^T)\widetilde{\mathbf{s}}$.

I2DUPEN Algorithm. Let now p_μ, c_μ denote the local slope and curvature values of the 2D distribution point μ and I_i be the indices subset related to the neighborhood of the distribution point i (see [2] for detailed explanation), the I2DUPEN algorithm is stated as follows:

(i) Compute $\widetilde{\mathbf{K}}_2 = \mathbf{B}\overline{\mathbf{K}}_2$.

(ii) Compute $\widetilde{\mathbf{s}} = (\mathbf{B} \otimes \mathbf{I})\overline{\mathbf{s}}$.

(iii) Compute the TSVD of $\mathbf{K}_1$ and $\widetilde{\mathbf{K}}_2$ and set $\hat{\mathbf{K}}_1 = \hat{\mathbf{\Sigma}}_1\hat{\mathbf{V}}_1^T$, $\hat{\mathbf{K}}_2 = \hat{\mathbf{\Sigma}}_2\hat{\mathbf{V}}_2^T$.

(iv) Compute $\hat{\mathbf{s}} = (\hat{\mathbf{U}}_2^T \otimes \hat{\mathbf{U}}_1^T)\widetilde{\mathbf{s}}$.

(v) Set $k = 0$ and compute an over smoothed solution $\mathbf{f}^{(k)}$ of the problem

$$\min_{\mathbf{f}\geq 0} \|(\hat{\mathbf{K}}_2 \otimes \hat{\mathbf{K}}_1)\mathbf{f} - \hat{\mathbf{s}}\|^2$$

by applying a few iterations of the Gradient Projection method [10].

(vi) **repeat**

(a) Compute $\lambda_i^{(k)} = \dfrac{\|(\hat{\mathbf{K}}_2 \otimes \hat{\mathbf{K}}_1)\mathbf{f}^{(k)} - \hat{\mathbf{s}}\|^2}{N\left(\beta_0 + \beta_p \max_{\mu\in I_i}(p_\mu^{(k)})^2 + \beta_c \max_{\mu\in I_i}(c_\mu^{(k)})^2\right)}, \quad i = 1, \ldots, N.$

(b) Compute $\mathbf{f}^{(k+1)}$ by solving

$$\min_{\mathbf{f}\geq 0} \|(\hat{\mathbf{K}}_2 \otimes \hat{\mathbf{K}}_1)\mathbf{f} - \hat{\mathbf{s}}\|^2 + \sum_{i=1}^{N} \lambda_i^{(k)}(\mathbf{L}\mathbf{f})_i^2$$

using the Newton Projection method.

until $\|\mathbf{f}^{(k+1)} - \mathbf{f}^{(k)}\| \leq tol\ \|\mathbf{f}^{(k)}\|$

where $\beta_0, \beta_p, \beta_c$ are 2DUPEN parameters [2]. We observe that the core iteration of 2DUPEN is preserved in I2DUPEN and the filters act only on the kernels and the data. The SVDs of the kernel matrices are performed only once and their computation cost do not affect significantly the global computation cost.

4. Numerical Results

The effectiveness of the proposed strategies is extensively evaluated on a synthetic test problem in paragraph 4.1 and its behaviour on real data is described in paragraph 4.2.

4.1. Experiments on a synthetic test problem

In this paragraph we compare, on a synthetic test problem, the proposed improvement strategies 2DUPENm$_1$, 2DUPENm$_2$, 2DUPENm$_3$ and I2DUPEN. The following parameters have been used in all the 2DUPEN modifications: $\beta_0 = 10^{-6}$, $\beta_p = \beta_c = 1$, $tol = 0.001$. We consider here a model distribution $\mathbf{f}^* \in \mathbb{R}^N$, $N = 64 \cdot 64$, and use it to synthesize the simulated IR-CPMG sequence data with 128×2048 non uniformly distributed $T_1 - T_2$ data. The test problem has two Gaussian peaks positioned along the main diagonal over a zero flat area (figure 1(a)). The noisy data $\mathbf{s}$ are obtained by adding Gaussian white noise of level δ to the synthetic data: $\mathbf{s} = (\mathbf{K}_2 \otimes \mathbf{K}_1)\mathbf{f}^* + \mathbf{e}$ where $\mathbf{e} = \delta\boldsymbol{\eta}$ and $\boldsymbol{\eta}$ is a normal random Gaussian vector such that $\|\boldsymbol{\eta}\| = 1$. Several values of δ have been tested, we report here the results obtained with $\delta = 10^{-2}$ which is close to the noise level in the real data. The Relative Error (Err) is used to evaluate the reconstruction quality. In table 1 we report the reconstruction errors and the computation time for the 2DUPENm$_1$, 2DUPENm$_2$ and I2DUPEN algorithms for different threshold values τ. The 2DUPEN algorithm solves this problem in 17770.81 s with relative error Err $= 1.972\ 10^{-2}$ (figure 1(b)). Observing the left picture of figure 2 we see the log-uniform decreasing distribution

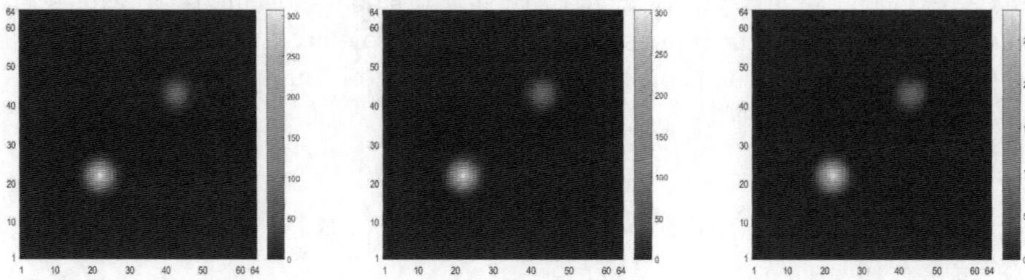

(a) Map of the model distribution $\mathbf{f}^*$ (b) 2DUPEN (c) I2DUPEN

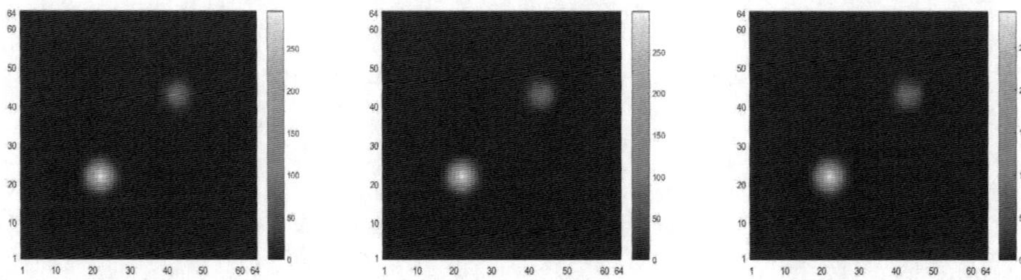

(d) 2DUPENm$_1$ (e) 2DUPENm$_2$ (f) 2DUPENm$_3$

Figure 1: Model distribution $\mathbf{f}^*$ and Reconstructed 2D maps. ($\tau = 10^{-2}$ for (c), (d), (e)).

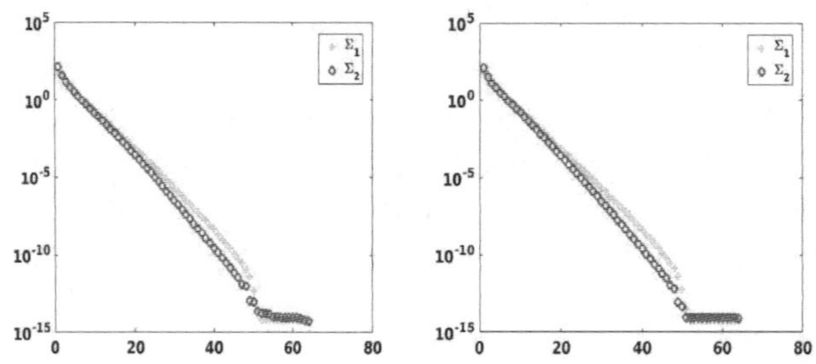

Figure 2: Singular values of $\mathbf{K}_1$ and $\mathbf{K}_2$ (left) and $\mathbf{K}_1$ and $\widetilde{\mathbf{K}}_2$ (right).

of the singular values of $\mathbf{K}_1$ and $\mathbf{K}_2$ yielding to a severe ill conditioning of the problem. At first we introduce only the SVD filter with threshold τ ranging from 0 (i.e. no filter) to 10^{-2} (2DUPENm$_1$). In this case, the whole acquisition set of $128 \cdot 2048$ data is reduced by projection to $64 \cdot 64$ samples when $\tau = 0$ down to a minimum of $12 \cdot 11$ samples when $\tau = 10^{-1}$, as reported in column 2 of table 1. We observe that the size reduction causes a proportional decrease of the computational time and an increase of the relative error values (columns 3 and 4 of table 1). The computed distribution is shown in figure 1(d). We examine now the results obtained using data reduced by the MW filter. In our example the reduced data have $128 \cdot 118$ samples. The singular values distribution does not change significantly for the introduction of the weights (right plot of figure 2) and the problem is still badly ill-conditioned. We see that the MW filter causes an increases of the reconstruction errors (columns 6 and 9 of table 1) even if the computed 2D maps are visually indistinguishable. We observe that I2DUPEN has always the smallest computation times and gives restorations with comparable qualitative accuracy. Finally, 2DUPENm$_3$ obtains, in 2062.90 s, a computed 2D map with relative error $Err = 5.002 \cdot 10^{-2}$ (figure 1(f)).

	2DUPENm$_1$			2DUPENm$_2$			I2DUPEN		
τ	ρ_1, ρ_2	Err	Time(s)	ρ_1, ρ_2	Err	Time(s)	ρ_1, ρ_2	Err	Time(s)
0	64, 64	2.749e-02	1352.93	64, 64	3.717e-02	2444.87	64, 64	7.402e-02	925.04
10^{-6}	32, 29	2.620e-02	1455.08	32, 28	3.768e-02	1598.35	32, 29	5.571e-02	884.07
10^{-4}	24, 22	2.731e-02	1252.22	24, 21	3.827e-02	1362.99	24, 22	6.124e-02	840.35
10^{-3}	20, 18	2.841e-02	907.54	20, 17	3.878e-02	1409.33	20, 18	5.655e-02	915.50
10^{-2}	16, 15	3.245e-02	839.48	16, 14	3.914e-02	1295.50	16, 15	5.833e-02	826.31
10^{-1}	12, 11	4.459e-02	738.53	12, 10	5.214e-02	778.53	12, 11	6.706e-02	720.71

Table 1: Results of 2DUPENm$_1$, 2DUPENm$_2$ and I2DUPEN algorithms on synthetic data.

From the different reconstructions and numerical results reported in figure 1 and table 1, we observe that the introduction of a filtering technique causes an increase of the reconstruction errors. We observe that when the full data $\mathbf{s} \in \mathbb{R}^M$ are available, 2DUPENm$_1$ is the best improvement of 2DUPEN both in terms of achieved quality and time reduction for large values of τ. However, if only the windowed data $\bar{\mathbf{s}}$ are available, 2DUPENm$_2$ is the best way to improve 2DUPEN accuracy, while I2DUPEN is the best improvement of 2DUPEN computational efficiency. Concluding, if the whole data set is available it is not convenient to apply MW because 2DUPENm$_1$ with large threshold (for example $\tau \simeq 10^{-3}$) efficiently computes the most accurate distribution. On the other hand, when the data are available only in MW compressed form, I2DUPEN is more suitable because it is faster than 2DUPENm$_2$ even if slightly less accurate.

4.2. Experiments on real data

In this paragraph we report the results obtained by testing I2DUPEN and its modifications on a real data set. A SR-CPMG test set of moderate size is used in order to be able to test 2DUPEN and all the I2DUPEN variants. A sample ($5 \times 5 \times 2$ cm^3) of biocalcarenite with porosity about 47%, was saturated under high vacuum with distilled water and promptly sealed into glass containers in order to avoid the evaporation of water. NMR measurements were performed at 25 ℃ by a NMR single-sided apparatus based on a Kea II NMR console and a NMR-MOUSE PM25 (Magritek, GmbH), controlled by the software Prospa (Magritek, GmbH). The 2D measurement, longitudinal-transverse relaxation curve ($T_1 - T_2$) was acquired by a SR-CPMG pulse sequence. The pulse sequence is characterized by an application of a train of $\pi/2$ pulses, in order to pre-saturate the transverse magnetization, and an encoding-reading stage $((\pi/2)_x - \tau - (\pi/2)_x - T_E/2 - [\pi_y - T_E/2 - \text{echo acquisition} - T_E/2]N_E)$. The T_1 relaxation signal was acquired with 32 values of τ chosen in geometrical progression from 1 ms up to 400 ms, with $N_E = 1024$ (number of acquired echos, echo times $T_E = 60\mu s$) on each CPMG, and number of scans equal to 24. All curves were acquired using phase-cycling procedures. The following parameters have been used in all the 2DUPEN modifications: $\beta_0 = 10^{-5}$, $\beta_p = 10, \beta_c = 30$, $tol = 0.05$. The reconstruction with $N_1 \times N_2 = 100 \times 100$ points has been computed by 2DUPEN in 72.55 s. The 3D surface and contour map are reported in figure 3: a peak hight $1.25 \cdot 10^{-5}$ a.u. (arbitrary unit) is located at $T_2 = 53$ ms and $T_1 = 135$ ms. Small peaks closed to the bottom border, probably are related to some small artefacts on the acquired data. It is noteworthy to highlight that the UPEN algorithm it isn't only a powerful tool to invert data, but also a tool to evaluate the quality of acquired data.

In this case the data reduced by MW have 32×360 samples and the diagonal weights matrix $\mathbf{B}$ has size 360×360.

The computation times obtained by 2DUPENm$_1$, 2DUPENm$_2$ and I2DUPEN with different threshold values are reported in table 2 together with the position and heigth of the main

peak. The ρ_1, ρ_2 values are the same for all the methods. As for synthetic data, 2DUPENm$_2$ is the less performing algorithm since always gives an inaccurate localization of the peak even if the computation times are comparable to those of I2DUPEN. We remark that the correct peaks localization is a very important information to be recovered in real applications. Concerning 2DUPENm$_1$ it requires the highest computation times and obtains a quite precise peaks localization. Hence we can conclude that I2DUPEN is the best improvement of the 2DUPEN algorithm for accuracy and computation times.
As an example we report the contour plots, maps and surface obtained by I2DUPEN, I2DUPENm$_1$ and I2DUPENm$_2$ with $\tau = 10^{-5}$ (figures 4, 5 and 6 respectively).

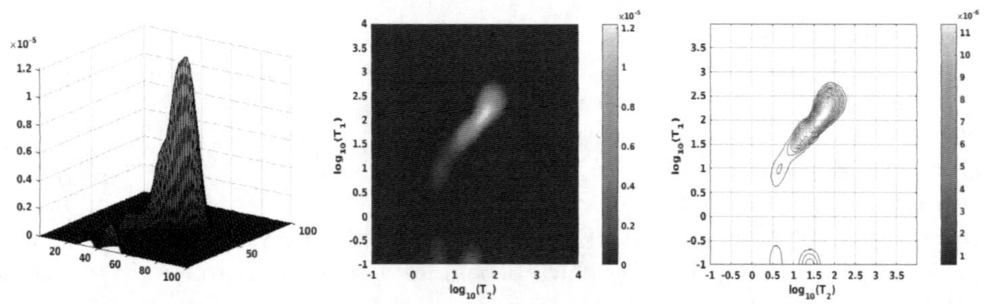

Figure 3: 3D surface (left), map (center) and contour (right) obtained by 2DUPEN.

		2DUPENm$_1$			2DUPENm$_2$			I2DUPEN		
τ	ρ_1, ρ_2	T_2, T_1	Height	Time	T_2, T_1	Height	Time	T_2, T_1	Height	Time
10^{-16}	32,100	53,120	2.22e-5	131.34	34,76	1.14e-5	44.01	53,121	2.09e-5	56.89
10^{-10}	25,32	53,135	1.39e-5	81.44	38,95	1.17e-5	50.76	53,135	1.39e-5	52.06
10^{-8}	22, 28	53,135	1.36e-5	93.25	38,95	1.13e-5	35.80	53,135	1.46e-5	42.08
10^{-5}	17, 21	48,120	1.31e-5	68.89	34,85	1.03e-5	36.41	53,135	1.31e-5	39.39
10^{-2}	10, 13	53,135	1.26e-5	64.82	38,96	1.01e-5	42.97	53,135	1.26e-5	36.10

Table 2: Real data. Results of 2DUPENm$_1$, 2DUPENm$_2$ and I2DUPEN. T_2, T_1 in milliseconds, Time in seconds and Height in arbitrary unit.

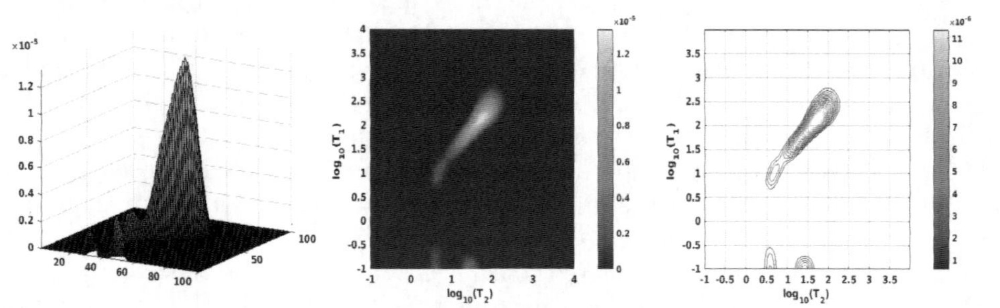

Figure 4: 3D surface (left), map (center) and contour (right) obtained by I2DUPEN with $\tau = 10^{-5}$. T_2, T_1 in milliseconds.

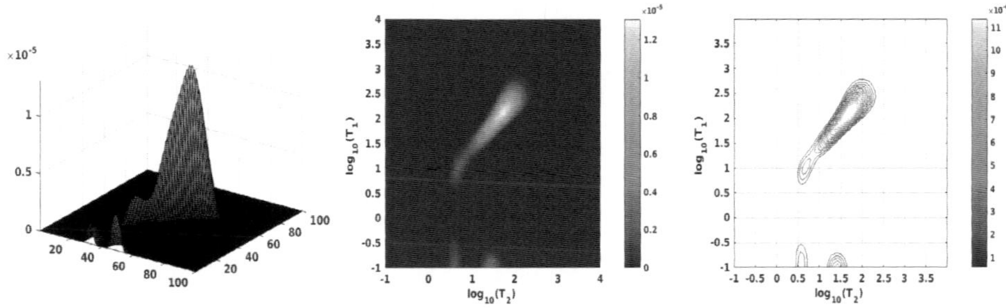

Figure 5: 3D surface (left), map (center) and contour (right) obtained by I2DUPENm$_1$ with $\tau = 10^{-5}$. T_2, T_1 in milliseconds.

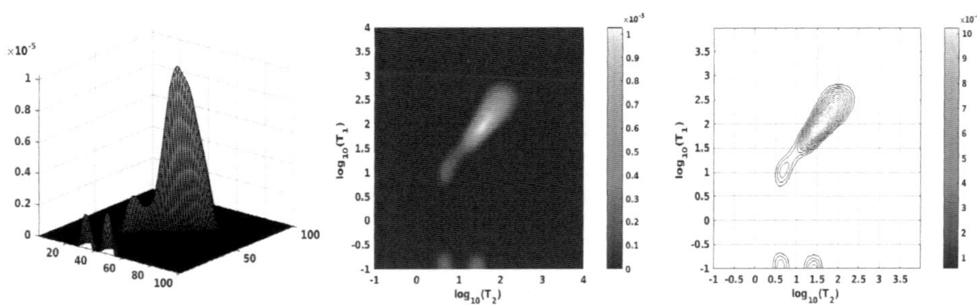

Figure 6: 3D surface (left), map (center) and contour (right) obtained by I2DUPENm$_2$ with $\tau = 10^{-5}$. T_2, T_1 in milliseconds.

5. Conclusions

In several NMR applications, users are interested in efficient methods able to reconstruct relaxation time distributions where the peaks are correctly localized and the underlying 3D distribution volume is well approximated. Our experimental analysis shows that I2DUPEN is the best modification of 2DUPEN in order to efficiently obtain correct information about the unknown relaxation times distributions. The loss of information due to the MW filter is less critical on real data because of unavoidable instrumental limitations and artifacts. This explains the better accuracy of I2DUPEN compared to I2DUPENm$_1$ in case of real data.

Acknowledgments

The authors thank the institute ITMC - RWTH Aachen University and Sabina Haber-Pohlmeier for the assistance in 2D NMR measurements.

This work was partially supported by INdAM-GNCS, under the 2016 project *Nonlinear numerical methods for inverse problems and applications.*

References

[1] Song, Y.-Q. 2013 *Journal of Magnetic Resonance (San Diego, Calif.: 1997), 229, 1224*
[2] Bortolotti V, Brown R J S, Fantazzini P, Landi G and Zama F 2016 *Inverse Problems* **33** 015033
[3] Borgia G C, Brown R J S and Fantazzini P 2000 *Journal of Magnetic Resonance* **147** 273–285
[4] Bertsekas D 1999 *Nonlinear Programming* (Athena Scientific, (2nd Edition))
[5] Sezginer A 1994 *U.S. Patent 5,363,041*
[6] Venkataramanan L, Yi-Qiao S and Hurlimann M 2002 *IEEE Trans. Signal Processing* **50** 1017–1026
[7] Butler J. P., Reeds J. A., Dawson S. V., 1981 *SIAM Journal on Numerical Analysis* **18** 381-397
[8] Bortolotti V, Brown R J S, Fantazzini P, Landi G and Zama F 2017 *Microporous and Mesoporous Materials* doi: 10.1016/j.micromeso.2017.04.038

IOP Conf. Series: Journal of Physics: Conf. Series **904** (2017) 012005 doi:10.1088/1742-6596/904/1/012005

[9] Blümich B 2005 *Essential NMR* (Springer-Verlag)
[10] Cornelio A, Porta F, Prato M and Zanni L 2013 *Inverse Problems* **29** 125013

Efficient Smoothed Concomitant Lasso Estimation for High Dimensional Regression

Eugene Ndiaye†, Olivier Fercoq†, Alexandre Gramfort†, Vincent Leclère*, and Joseph Salmon†

†LTCI, Télécom ParisTech, Université Paris-Saclay, 46 rue Barrault, 75013 Paris, France
*Université Paris-Est, Cermics (ENPC), 77455 Marne-la-Vallée, France

Abstract.
In high dimensional settings, sparse structures are crucial for efficiency, both in term of memory, computation and performance. It is customary to consider ℓ_1 penalty to enforce sparsity in such scenarios. Sparsity enforcing methods, the Lasso being a canonical example, are popular candidates to address high dimension. For efficiency, they rely on tuning a parameter trading data fitting versus sparsity. For the Lasso theory to hold this tuning parameter should be proportional to the noise level, yet the latter is often unknown in practice. A possible remedy is to jointly optimize over the regression parameter as well as over the noise level. This has been considered under several names in the literature: Scaled-Lasso, Square-root Lasso, Concomitant Lasso estimation for instance, and could be of interest for uncertainty quantification. In this work, after illustrating numerical difficulties for the Concomitant Lasso formulation, we propose a modification we coined Smoothed Concomitant Lasso, aimed at increasing numerical stability. We propose an efficient and accurate solver leading to a computational cost no more expensive than the one for the Lasso. We leverage on standard ingredients behind the success of fast Lasso solvers: a coordinate descent algorithm, combined with safe screening rules to achieve speed efficiency, by eliminating early irrelevant features.

1. Introduction

In the context of high dimensional regression where the number of features is greater than the number of observations, standard least squares need some regularization to both avoid over-fitting and ease the interpretation of discriminant features. Among the least squares with sparsity inducing regularization, the Lasso [27], using the ℓ_1 norm as a regularizer, is the most standard one. It hinges on a regularization parameter governing the trade-off between data fitting and sparsity of the estimator, and requires careful tuning. Though this estimator is well understood theoretically, the choice of the tuning parameter remains an open and critical question in practice as well as in theory. For the Lasso, statistical guarantees [5] rely on choosing the tuning parameter proportional to the noise level, a quantity that is usually unknown. Besides, the noise level is of practical interest since it is required in the computation of model selection criterions such as AIC, BIC, SURE or in the construction of confidence sets. A convenient way to estimate both the regression coefficient and the noise level is to perform a joint estimation.

The most famous road for this joint estimation was inspired by the robust theory developed by Huber [14], particularly in the context of location-scale estimation. Indeed, Owen [20] extended it to handle sparsity inducing penalty, leading to a jointly convex optimization formulation. Since then, his estimator has appeared under various name, and we coined it the Concomitant Lasso.

IOP Conf. Series: Journal of Physics: Conf. Series **904** (2017) 012006 doi:10.1088/1742-6596/904/1/012006

Later, the same formulation was mentioned in [1], in a response to [23], and was thoroughly analyzed in [25], under the name Scaled-Lasso. Similar results were independently obtained in [4] for the same estimator, though with a different formulation. While investigating pivotal quantities, Belloni *et al* proposed to solve the following convex program: modify the standard Lasso by removing the square in the data fitting term. Thus, they termed their estimator the Square-root Lasso (see also [6]). A second approach leading to this formulation was also proposed by [28] to account for noise in the design matrix in an adversarial scenario. Interestingly their robust construction led exactly to the Square-root Lasso formulation.

Under standard design assumption (see [5]), it is proved that the Scaled/Square-root Lasso reaches optimal rates for sparse regression, with the additional benefit that the regularization parameter is independent of the noise level [4, 25]. Moreover, a practical study [22] has shown that the Concomitant Lasso estimator, or its debiased version (see for instance [3, 15] for a discussion on least-squares refitting), is particularly well suited for estimating the noise level.

Among the solutions to compute the Concomitant Lasso, two roads have been explored. Considering the Scaled-Lasso formulation, Sun and Zhang [24, 25] have proposed an iterative procedure that alternates Lasso and noise estimation steps, the later leading to rescaling the tuning parameter iteratively. On the other hand, considering the Square-root Lasso formulation, Belloni *et al* [4] have leaned on second order cone programming solvers, *e.g.,* TFOCS [2].

Despite the appealing properties listed above, among which the superiority of the theoretical results is the most striking, no consensus for an efficient solver has yet emerged. Our contribution aims at providing a more numerically stable formulation, called the Smoothed Concomitant Lasso. This variant allows to obtain a fast solver based on coordinate descent. Then, we propose dedicated safe rules, as introduced in [9, 12] for the Lasso. We show similar accelerations for the Smoothed Concomitant Lasso, both on real and simulated data. Overall, our method presents the same computational cost as for the Lasso, but enjoys the nice features mentioned earlier in terms of statistical properties.

2. Concomitant estimator

Next we present our estimator and some important properties. Proofs are in [17, Appendix].

Notation For any integer $d \in \mathbb{N}$, we denote by $[d]$ the set $\{1, \dots, d\}$. Our observation vector is $y \in \mathbb{R}^n$ (assumed to be nonzero) and the design matrix $X = [X_1, \dots, X_p] \in \mathbb{R}^{n \times p}$ has p explanatory variables or features. The Euclidean norm is written $\|\cdot\|$, the ℓ_1 norm $\|\cdot\|_1$, the ℓ_∞ norm $\|\cdot\|_\infty$, and the matrix transposition of a matrix Q is denoted by $Q^\top$. We note $\mathcal{B}_\infty$ the unit ball with the ℓ_∞ norm. For real numbers a and b, $a \vee b$ stands for the maximum of a and b. We denote $\mathcal{S}_\tau$ the soft-thresholding operator at level $\tau > 0$, $\mathcal{S}_\tau(x) = \operatorname{sign}(x)(|x| - \tau)_+$. For a closed convex set $\mathcal{C}$, we write $\Pi_\mathcal{C}$ the projection. The sub-gradient of a convex function $f : \mathbb{R}^d \to \mathbb{R}$ at x is defined as $\partial f(x) = \{z \in \mathbb{R}^d : \forall y \in \mathbb{R}^d, f(x) - f(y) \geq z^\top(x - y)\}$. We denote by ι_C the indicator function of a set C defined as $\iota_C(x) = 0$ if $x \in C$ and $\iota_C(x) = \infty$ if $x \notin C$. We recall that the sub-differential $\partial\|\cdot\|_1$ of the ℓ_1 norm is the set-valued function $\operatorname{sign}(\cdot)$, defined element-wise for all $j \in [d]$ by $\operatorname{sign}(x_j) = 1$ if $x_j > 0$, by $\operatorname{sign}(x_j) = -1$ if $x_j < 0$ and by $\operatorname{sign}(x_j) = [-1, 1]$ if $x_j = 0$. For a set $S \subset [p]$, we denote by $P_{X,S} = X_S \left(X_S^\top X_S\right)^+ X_S^\top$ the projection operator onto $\operatorname{Span}\{X_j : j \in S\}$, where A^+ represents the Moore-Penrose pseudo-inverse. We note tr(X) the trace of matrix X and $\widehat{\Sigma} = X^\top X/n$ the normalized Gram matrix of X.

2.1. *Concomitant Lasso*

Let us first introduce the Concomitant Lasso estimator, following the formulation proposed in [20, 25], and present some properties obtained due to convexity and duality.

Definition 1. For $\lambda > 0$, the Concomitant Lasso estimator $\hat{\beta}^{(\lambda)}$ is defined as a solution of the primal optimization problem

$$(\hat{\beta}^{(\lambda)}, \hat{\sigma}^{(\lambda)}) \in \mathop{\arg\min}_{\beta \in \mathbb{R}^p, \sigma > 0} P_\lambda(\beta, \sigma) := \frac{\|y - X\beta\|^2}{2n\sigma} + \frac{\sigma}{2} + \lambda \|\beta\|_1 . \quad (1)$$

Theorem 1. *Denoting* $\Delta_{X,\lambda} = \left\{\theta \in \mathbb{R}^n : \|X^\top \theta\|_\infty \leq 1, \lambda\sqrt{n}\|\theta\| \leq 1\right\}$, *the dual formulation of the Concomitant Lasso reads*

$$\hat{\theta}^{(\lambda)} \in \mathop{\arg\max}_{\theta \in \Delta_{X,\lambda}} D_\lambda(\theta) := \langle y, \lambda\theta \rangle . \quad (2)$$

For an optimal primal vector $\hat{\beta}^{(\lambda)}$, $\hat{\sigma}^{(\lambda)} = \|y - X\hat{\beta}^{(\lambda)}\|/\sqrt{n}$. *Moreover, Fermat's rule reads as the link-equation between primal and dual solutions* $y = n\lambda\hat{\sigma}^{(\lambda)}\hat{\theta}^{(\lambda)} + X\hat{\beta}^{(\lambda)}$ *and the sub-differential inclusion* $X^\top(y - X\hat{\beta}^{(\lambda)}) \in n\lambda\hat{\sigma}^{(\lambda)}\partial \|\cdot\|_1 (\hat{\beta}^{(\lambda)})$.

As defined in (1), the Concomitant Lasso estimator is ill-defined: the set over which we optimize is not closed and no solution may exist. We circumvent this difficulty by considering instead the Fenchel biconjugate of the objective function (for more details, see [17, Appendix C]). The actual objective function accepts $\sigma \geq 0$ as soon as $y = X\beta$. In the rest of the paper, we will write (1) instead of the minimization of the biconjugate as a slight abuse of notation.

A principled way to estimators regularization parameters of Lasso-type programs is to use cross-validation over a fixed set of parameters. Usually, a geometrical grid $\lambda_t = \lambda_{\max}^{\mathrm{L}} 10^{-\delta(t-1)/(T-1)}, t \in [T]$ is used. `scikit-learn` [10] and `glmnet` [13] set $\delta = 3$.

2.2. Critical parameters for the Concomitant Lasso

As for the Lasso, the null vector is optimal for the Concomitant Lasso problem as soon as the regularization parameter becomes too large, as detailed in the next proposition.

Proposition 1. *We have* $\hat{\beta}^{(\lambda)} = 0$ *for all* $\lambda \geq \lambda_{\max} := \|X^\top y\|_\infty / (\|y\|\sqrt{n})$.

However, for the Concomitant Lasso, there is another extreme. Indeed, there exists a critical parameter $\lambda_{\min}$ such that the Concomitant Lasso is equivalent to the Basis Pursuit for all $\lambda \leq \lambda_{\min}$ and gives an estimate $\hat{\sigma}^{(\lambda)} = 0$. The Basis Pursuit and its dual are given by $\hat{\beta}^{\mathrm{BP}} \in \arg\min_{\beta \in \mathbb{R}^p : y = X\beta} \|\beta\|_1$ and $\hat{\theta}^{\mathrm{BP}} \in \arg\max_{\theta \in \mathbb{R}^n : \|X^\top\theta\|_\infty \leq 1} \langle y, \theta \rangle$.

Proposition 2. *For any* $\hat{\theta}^{\mathrm{BP}} \in \arg\max_{\theta \in \mathbb{R}^n : \|X^\top\theta\|_\infty \leq 1} \langle y, \theta \rangle$ *and any* $\lambda \leq \lambda_{\min} := 1/(\|\hat{\theta}^{\mathrm{BP}}\|\sqrt{n})$, $(\hat{\beta}^{\mathrm{BP}}, 0)$ *is optimal for* P_λ *and* $\hat{\theta}^{\mathrm{BP}}$ *is optimal for* D_λ.

2.3. Smoothed Concomitant Lasso

We can guarantee the existence of minimizers to the Concomitant Lasso (see [17, Appendix C]), even if $\hat{\sigma}^{(\lambda)} = 0$, but the problem becomes more and more ill-conditioned. The previous proposition shows that for too small λ's, a Basis Pursuit solution will always be found, though numerically this might be challenging to get. Indeed, when λ approaches $\lambda_{\min}$, a coordinate descent algorithm (similar to the one described in Algorithm 1) starts to become significantly slower. To avoid this issue, we propose a slight modification of the objective function by adding a constraint on σ, which corresponds to a noise level limit σ_0. We refer to this method as the Smoothed Concomitant Lasso following the terminology introduced by Nesterov in [19].

Definition 2. For $\lambda > 0$ and $\sigma_0 > 0$, the Smoothed Concomitant Lasso estimator $\hat{\beta}^{(\lambda,\sigma_0)}$ and its associated noise level estimate $\hat{\sigma}^{(\lambda,\sigma_0)}$ are defined as solutions of the primal optimization problem

$$\mathop{\arg\min}_{\beta \in \mathbb{R}^p, \sigma \in \mathbb{R}} P_{\lambda,\sigma_0}(\beta, \sigma) := \frac{\|y - X\beta\|^2}{2n\sigma} + \frac{\sigma}{2} + \lambda \|\beta\|_1 + \iota_{[\sigma_0, +\infty[}(\sigma). \quad (3)$$

Remark 1. Concurrently to our work, [16] proposed a similar smoothing strategy on the Square-root Lasso formulation, equivalent to our formulation. They also show in [16, Theorem 3.7] that the Smoothed Concomitant Lasso preserves the statistical consistency of the Concomitant Lasso: it achieves the minimax optimal rate of convergence as soon as $\lambda \geq 24\sqrt{\log(p)/n}$ and $\sigma_0 \leq \sigma/4$.

Theorem 2. *The dual formulation of the Smoothed Concomitant Lasso reads*

$$\hat{\theta}^{(\lambda,\sigma_0)} = \operatorname*{arg\,max}_{\theta \in \Delta_{X,\lambda}} D_{\lambda,\sigma_0}(\theta) := \langle y, \lambda\theta \rangle + \sigma_0 \left(\frac{1}{2} - \frac{\lambda^2 n}{2} \|\theta\|^2 \right), \tag{4}$$

for $\Delta_{X,\lambda} = \left\{\theta \in \mathbb{R}^n : \|X^\top \theta\|_\infty \leq 1, \|\theta\| \leq 1/(\lambda\sqrt{n})\right\}$. *Associated to an optimal* $\hat{\beta}^{(\lambda,\sigma_0)}$, *we must have* $\hat{\sigma}^{(\lambda,\sigma_0)} = \sigma_0 \vee (\|y - X\hat{\beta}^{(\lambda,\sigma_0)}\|/\sqrt{n})$. *Also, the link-equation* $y = n\lambda\hat{\sigma}^{(\lambda,\sigma_0)}\hat{\theta}^{(\lambda,\sigma_0)} + X\hat{\beta}^{(\lambda,\sigma_0)}$ *and the sub-differential inclusion* $X^\top(y - X\hat{\beta}^{(\lambda,\sigma_0)}) \in n\lambda\hat{\sigma}^{(\lambda,\sigma_0)}\partial\|\cdot\|_1(\hat{\beta}^{(\lambda,\sigma_0)})$ *hold.*

Remark 2. Since $D_{\lambda,\sigma_0}(\theta)$ is strongly concave and $\Delta_{X,\lambda}$ is convex and closed, $\hat{\theta}^{(\lambda,\sigma_0)}$ is unique.

In practice, the choice of σ_0 can be motivated as follows:

- Suppose we have prior information on the minimal noise level expected in the data. Then we can set σ_0 as this bound. Indeed, if $\hat{\sigma}^{(\lambda,\sigma_0)} > \sigma_0$, then the constraint $\sigma \geq \sigma_0$ is not active and the minimizers of Problem (3) are also minimizers of Problem (1). The Smoothed Concomitant Lasso estimator will only be different from the Concomitant Lasso estimator when the prediction given by the Concomitant Lasso violates the a priori information.
- Without prior information we can consider a given accuracy ε, and set $\sigma_0 = \varepsilon$. Then, the theory of smoothing [19] tells us that any $\varepsilon/2$-solution to Problem (3) is an ε-solution to Problem (1). Thus we obtain the same solutions, but as an additional benefit we have a control on the conditioning of the problem.
- If departing slightly from the Concomitant Lasso estimator is not too big of an issue, one can also use an arbitrary proportion of the initial estimation of the noise variance *i.e.*, $\sigma_0 = \|y\|/\sqrt{n} \times 10^{-\alpha}$. This was our choice in practice, and we have set $\alpha = 2$. Indeed, taking a large enough value for σ_0 leads to less numerical issues.

A similar reasoning to Proposition 1 gives the following critical parameter.

Proposition 3. *We have* $\hat{\beta}^{(\lambda,\sigma_0)} = 0$, *for all* $\lambda \geq \lambda_{\max} := \|X^\top y\|_\infty / (n\,(\sigma_0 \vee (\|y\|/\sqrt{n})))$.

2.4. Duality gap and link with the Lasso

By comparing Fermat's rule in Theorems 1 and 2, one can remark that if $\hat{\beta}^{(\lambda,\sigma_0)}$ is a solution of the Smoothed Concomitant Lasso, then it is also a solution of the Lasso with regularization parameter $\lambda\hat{\sigma}^{(\lambda,\sigma_0)}$. The following proposition estimates the quality (in term of duality gap) of a primal-dual vector in the Lasso path compared to Concomitant Lasso path. Denoting $G^{\mathrm{L}}_{\sigma\lambda}(\beta,\theta)$ the duality gap for the standard lasso, one can easily show that

Proposition 4. $\forall \beta \in \mathbb{R}^p, \theta \in \Delta_{X,\lambda}, \sigma \geq \sigma_0$, $G^{\mathrm{L}}_{\sigma\lambda}(\beta,\theta) \leq \sigma G_{\lambda,\sigma_0}(\beta,\sigma,\theta)$.

Hence, as $\forall\lambda$, $\hat{\sigma}^{(\lambda)} \leq \|y\|/\sqrt{n}$, if the duality gap for the Smoothed Concomitant Lasso is small, so is the duality gap for the Lasso with the corresponding regularization parameter.

3. Safe screening rules

In order to achieve greater computational efficiency, we propose new safe screening rules (using the terminology introduced in [9]) for our problem and we compare their performance. The principle underlying safe screening rules is as follows: one can discard inactive features from the optimization problem, thanks to the sub-differential inclusion in Theorem 2 and to a safe region $\mathcal{R}$ such that $\hat{\theta}^{(\lambda,\sigma_0)} \in \mathcal{R}$. Indeed if $\max_{\theta\in\mathcal{R}} |X_j^\top \theta| < 1$ then $|X_j^\top \hat{\theta}^{(\lambda,\sigma_0)}| < 1$ and thus $\hat{\beta}_j^{(\lambda,\sigma_0)} = 0$.

Algorithm 1: CD4SCL – Coordinate Descent for the Smoothed Concomitant Lasso with Gap Safe screening

```
Input  : X, y, ε, K, f^ce(= 10), λ, σ_0, β, σ
A ← [p]
for k ∈ [K] do
    if k mod f^ce = 1 then
        Compute θ as in Proposition 6
        if G_{λ,σ_0}(β, σ, θ) = P_{λ_t,σ_0}(β, σ) − D_{λ_t,σ_0}(θ) ≤ ε. then      // Stopping criterion
            break
        Update A thanks to Proposition 5                                   // Screening test
    for j ∈ A do                                                           // Loop over coordinates
        β_j ← S_{nσλ_t/||X_j||^2}(β_j − X_j^⊤(Xβ − y)/||X_j||^2)           // Soft-thresholding step
        σ ← σ_0 ∨ (||y − Xβ||/√n))                                         // Noise estimation step
Output: β, σ, A
```

Since the dual objective of the Smoothed Concomitant Lasso is $\lambda^2\sigma_0 n$-strongly concave, we can provide a dynamic and converging SAFE sphere region $\mathcal{R}$, following [18].

Proposition 5 (Gap Safe rule). *For all $(\beta, \sigma, \theta) \in \mathbb{R}^p \times \mathbb{R}_+ \times \Delta_{X,\lambda}$, then for $r = \sqrt{2G_{\lambda,\sigma_0}(\beta,\sigma,\theta)/(\lambda^2\sigma_0 n)}$, we have $\hat{\theta}^{(\lambda,\sigma_0)} \in \mathcal{B}(\theta, r)$. Thus, if $|X_j^\top\theta| + r\,\|X_j\| < 1$ then $\hat{\beta}_j^{(\lambda,\sigma_0)} = 0$.*

4. Algorithmic details

4.1. Smoothed Concomitant Lasso algorithm (SC)

We first present the inner loop of our main algorithm, *i.e.*, the implementation of coordinate descent for the Smoothed Concomitant Lasso. In Algorithm 1, we denote by $\mathcal{A}$ the active set, *i.e.*, the set of coordinates that we have not been screened out. For safe screening rules, this set is guaranteed to contain the support of the optimal solution.

The fast solver we used for the Smoothed Concomitant Lasso, rely on the two following key features: coordinate descent and Gap Safe screening rules.

4.2. Coordinate descent

The algorithm we consider to compute the Smoothed Concomitant Lasso is coordinate descent, an efficient way to solve Lasso-type problem (even for multiple values of parameters) [13]. Previous attempts mainly focused on iteratively alternating Lasso steps along with noise level estimation [25], or conic programming [2]. In [16], written concurrently to this work, the authors consider ISTA, a first order method using full gradient information at each iteration.

Here we provide a simple and efficient coordinate descent approach, *cf.* Algorithm 1. Our primal objective P_{λ,σ_0} can be written as the sum of a convex differentiable function $f(\beta,\sigma) = \|y - X\beta\|^2/(2n\sigma) + \sigma/2$ and of a separable function $g(\beta,\sigma) = \lambda\|\beta\|_1 + \iota_{[\sigma_0,+\infty[}(\sigma)$. Moreover, for $\sigma \geq \sigma_0 > 0$, the gradient of f is Lipschitz continuous. Hence, we know that the coordinate descent method converges to a minimizer of our problem [29]. We choose to update the variable σ every other iteration because this can be done at a negligible cost.

Our stopping criterion is based on the duality gap defined by $G_{\lambda,\sigma_0}(\beta,\sigma,\theta) = P_{\lambda,\sigma_0}(\beta,\sigma) - D_{\lambda,\sigma_0}(\theta)$. This requires the computation of a dual feasible point as follows.

Proposition 6. *Let $(\beta_k)_{k\in\mathbb{N}}$ be a sequence that converges to $\hat{\beta}^{(\lambda,\sigma_0)}$. Then $(\theta_k)_{k\in\mathbb{N}}$ built as $\theta_k = \frac{y - X\beta_k}{\lambda n\sigma_0 \vee \left\|X^\top(y - X\beta_k)\right\|_\infty \vee \lambda\sqrt{n}\|y - X\beta_k\|}$ converges to $\hat{\theta}^{(\lambda,\sigma_0)}$. Hence $G_{\lambda,\sigma_0}(\beta_k,\sigma_k,\theta_k) \to_{k\to+\infty} 0$.*

5. Numerical experiments

We compare the estimation performance and computation times of standard deviation estimators which are presently the state-of-the-art in high dimensional settings. We refer to [22] for a recent

$\hat{\sigma}_{OR}$	$\hat{\sigma}_{\mathcal{M}-CV}$	$\hat{\sigma}_{\mathcal{M}-LS}$	$\hat{\sigma}_i$	$\widehat{\sigma}_{D2}$
$\frac{\Vert y-P_{X,S^\star} y\Vert}{\sqrt{n-\vert S^\star\vert}}$	$\frac{\Vert y-X\hat{\beta}^{\lambda_{cv}}_{\mathcal{M}}\Vert}{\sqrt{n-\vert\hat{S}^{\lambda_{cv}}_{\mathcal{M}}\vert}}$	$\frac{\Vert y-P_{X,\hat{S}_{\mathcal{M}}} y\Vert}{\sqrt{n-\vert\hat{S}_{\mathcal{M}}\vert}}$	$\frac{\Vert y^{(i')}-P_{X^{(i')},\hat{S}_i} y^{(i')}\Vert}{\sqrt{n/2-\vert\hat{S}_i\vert}}$	$\frac{\left(1+\frac{p\widehat{m}_1^2}{(n+1)\widehat{m}_2}\right)\Vert y\Vert^2}{n} - \frac{\widehat{m}_1\Vert X^\top y\Vert^2}{\sqrt{n(n+1)\widehat{m}_2}}$

Table 1: The estimator $\hat{\beta}_{\mathcal{M}}$ are obtained by a method $\mathcal{M}$ and $\mathcal{M}-LS$ is its least square refitting. We note $S^{\odot} = \{j \in [p], \beta_j^{\odot} \neq 0\}$, $D_i = (y^{(i)}, X^{(i)})_{i\in[2]}$ is a split in two parts of the observations, and $\hat{S}_i$ the support selected after a cross-validation on the part D_i. The RCV estimator is $\hat{\sigma}_{RCV} = ((\hat{\sigma}_1^2 + \hat{\sigma}_2^2)/2)^{1/2}$, and $\widehat{m}_1 = \mathrm{tr}(\widehat{\Sigma})/p$ and $\widehat{m}_2 = \mathrm{tr}(\widehat{\Sigma}^2)/p - (\mathrm{tr}(\widehat{\Sigma}))^2/(pn)$.

comparison. In our simulations[1] we use the common setup: $y = X\beta^\star + \sigma\varepsilon$ where $\varepsilon \sim \mathcal{N}(0, \mathrm{Id}_n)$ and $X \in \mathbb{R}^{n\times p}$ follows a multivariate normal distribution with covariance $\Sigma = (\rho^{|i-j|})_{i,j\in[p]}$. We define $\beta^\star = \alpha\beta$ where the coordinates of β are drawn from a standard Laplace distribution and we randomly set $s\%$ of them to zero. The scalar α is chosen in order to satisfy a prescribed signal to noise ratio denoted snr: $\alpha = \sqrt{\mathrm{snr} \times \sigma^2/\beta^\top\Sigma\beta}$.

The procedures we have compared are summarized in Table 1. Namely, our reference is the oracle estimator (OR) $\hat{\sigma}_{OR}$, the cross-validated estimator (CV) $\hat{\sigma}_{\mathcal{M}-CV}$ whith a parameter λ_{cv} chosen by 5-fold cross-validation, the least-square refitting estimator (LS) $\hat{\sigma}_{\mathcal{M}-LS}$, the refitted cross-validation (RCV) $\hat{\sigma}_{RCV}$ and $\widehat{\sigma}_{D2}$ the estimator introduced in [8].

We run all the following algorithms over the non-increasing sequence $\lambda_t = \lambda_{\max}10^{-\delta(t-1)/(T-1)}, t \in [T]$ with the default value $\delta = 2, T = 100$. The regularization grid for the joint estimations (Scaled-Lasso, with solver from [25] (SZ), Smoothed Concomitant Lasso (SC), Square-root Lasso [4] (SQRT-Lasso) and the estimator introduced in [23] (SBvG)) begins at $\lambda_{\max}$ given in Proposition 3. We set Smoothed Concomitant Lasso with the default value $\sigma_0 = \Vert y\Vert/\sqrt{n} \times 10^{-2}$. As explained in Section 2.3 this choice improves numerical efficiency at the cost of departing slightly from the Concomitant Lasso estimator in the low noise regime. The grid for the Lasso (L) estimators begins with $\lambda_{\max}^{\mathrm{L}} = \Vert X^\top y\Vert_\infty/n$. The Lasso with the universal parameter $\lambda = \sqrt{2\log(p)/n}$ is denoted (L_U) and SZ refers to Concomitant Lasso with the quantile regularization described in [26] in Fig. 1(c).

5.1. Computational performance

Figure 1(a) presents on the Leukemia dataset the computation times observed for the different CV methods. The Smoothed Concomitant Lasso is based on the coordinate descent algorithm described in this paper and written in Python and Cython to generate low level C code, offering high performance. When a Lasso solver is needed, we have used the one from `scikit-learn`, that is coded similarly. For SZ_CV, computations are quite heavy as one uses the alternating algorithm proposed in [25]. Depending on the regularization parameter (for instance when one approaches $\lambda_{\min}$) the SZ_CV method is quite intractable and the algorithm faces the numerical issues mentioned earlier. The generic solver used for SBvG and SQRT-Lasso, is the `CVXPY` package [7], explaining why these methods are two orders of magnitude slower than a Lasso. In contrast, our solver reaches similar computing time w.r.t. an efficient Lasso solver, with the additional benefit of jointly estimating the coefficients and the standard deviation of the noise.

Figure 1(b) shows the benefit one can obtain thanks to the safe screening rule introduced above. Indeed, the Gap Safe rule greatly benefits from the convergence of the dual vector, leading to smaller and smaller safe sphere as the iterations proceeds [12, 18].

5.2. Performance of standard deviation estimators

As noted earlier in [11], spurious correlations can strongly affect sparse regression and usually lead to large biases. This makes the standard deviation estimation very challenging and affects

[1] Source code available at `https://github.com/EugeneNdiaye/smoothed_concomitant_lasso`

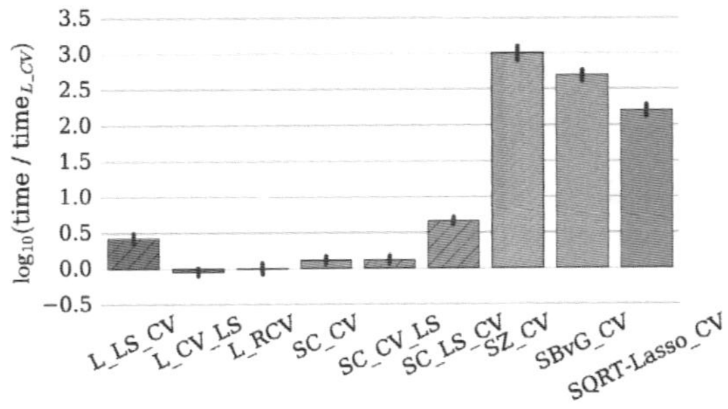

(a) Times to run simulations using synthetic dataset.

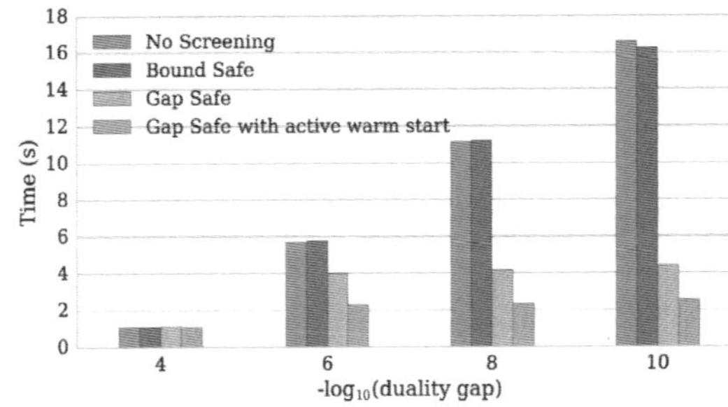

(b) Time to reach convergence using Leukemia dataset.

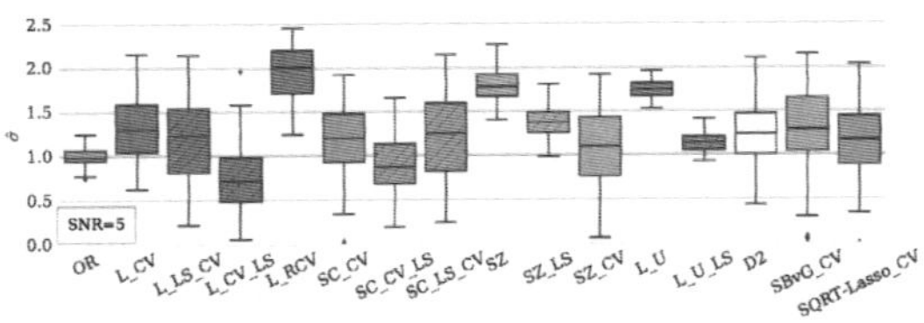

(c) Estimated performance using synthetic dataset.

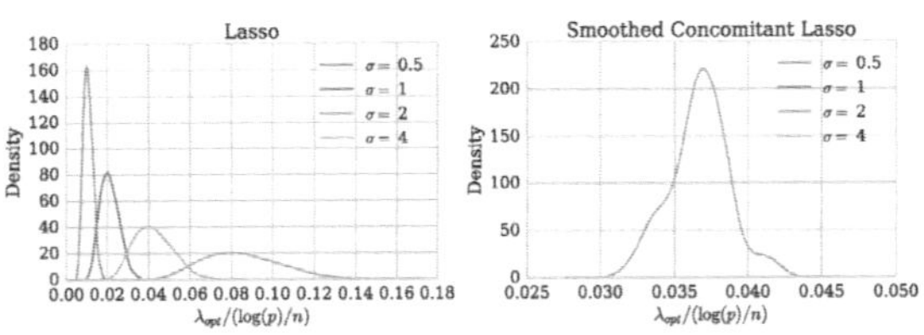

(d) Estimated distribution of the optimal λ_{opt}.

Figure 1: (a): comparisons of the computational times using different estimation method (time presented relative to the mean time of the Lasso). (b): speed up using screening rules for the Smoothed Concomitant Lasso w.r.t. to duality gap and for $(\lambda_t)_{t\in[100]}$. The dimensions of Leukemia dataset are $(n = 72, p = 7129)$. (c): comparison of quality of different estimators of the noise σ normalized to 1. The synthetic datasets are generated with the settings $(n = 100, p = 500, \rho = 0.6, \text{snr} = 5, s = 0.9, 50 \text{ replications})$. (d): comparisons of the distribution of optimal regularizer λ_{opt} under different levels of noise.

the cross-validation estimator based on the Lasso as they usually underestimate the standard deviation. The phenomenon is amplified when one uses least squares refitting on the cross-validated Lasso, as noticed in [22]. Results are presented as boxplots in Fig. 1(c) (see [17, Appendix B] for additional settings).

We have observed that SC and SZ are efficient in high sparsity settings with low correlations, correcting for the positive bias of the estimator from [23] (SBvG). In [22], it was also argued that a cross-validation estimator based on Lasso is more stable and performs better when the sparsity decreases and when the snr increases. Note that this is not the case when performing cross-validation with the Concomitant Lasso. Here, we show that the latter achieves similar performance as the Lasso. It is worth noting that our method is consistently good over the whole experiments we conducted especially when applying least squares refitting.

Another appealing property of the Smoothed Concomitant Lasso compared to the Lasso is the invariance of the optimal $\lambda_{opt} := \arg\min_{\lambda\in(\lambda_t)_{t\in[T]}} \|X\hat{\beta}^{(\lambda,\sigma_0)} - X\beta^\star\|_2$ w.r.t. different levels of noise. We show on Fig. 1(d) a kernel density plot of its distribution on synthetic data with different values of σ. A similar experiment was performed in [16] leading to the same conclusion with an optimal λ chosen by a train/test procedure.

6. Conclusion

We have explored the joint estimation of the coefficients and noise level for ℓ_1 regularized regression. We have corrected some numerical drawbacks of the Concomitant Lasso estimator by proposing a smoother formulation, leading to the Smoothed Concomitant Lasso. A fast algorithm, combining coordinate descent and safe screening rules was investigated achieved the same numerical efficiency than for the Lasso while estimating the noise level. Future research would extend our work to general data-fitting terms [20], and combine sketching techniques [21].

References

[1] A. Antoniadis. Comments on: ℓ_1-penalization for mixture regression models. *TEST*, 19(2):257–258, 2010.
[2] S. R. Becker, E. J. Candès, and M. C. Grant. Templates for convex cone problems with applications to sparse signal recovery. *Math. Program. Comput.*, 3(3):165–218, 2011.
[3] A. Belloni and V. Chernozhukov. Least squares after model selection in high-dimensional sparse models. *Bernoulli*, 19(2):521–547, 2013.
[4] A. Belloni, V. Chernozhukov, and L. Wang. Square-root Lasso: Pivotal recovery of sparse signals via conic programming. *Biometrika*, 98(4):791–806, 2011.
[5] P. J. Bickel, Y. Ritov, and A. B. Tsybakov. Simultaneous analysis of Lasso and Dantzig selector. *Ann. Statist.*, 37(4):1705–1732, 2009.
[6] S. Chrétien and S. Darses. Sparse recovery with unknown variance: a lasso-type approach. *IEEE Trans. Inf. Theory*, 2011.
[7] S. Diamond and S. Boyd. CVXPY: A Python-embedded modeling language for convex optimization. *J. Mach. Learn. Res.*, 2016. To appear.
[8] L. Dicker. Variance estimation in high-dimensional linear models. *Biometrika*, 101(2):269–284, 2014.
[9] L. El Ghaoui, V. Viallon, and T. Rabbani. Safe feature elimination in sparse supervised learning. *J. Pacific Optim.*, 8(4):667–698, 2012.
[10] F. Pedregosa *et al.* Scikit-learn: Machine learning in Python. *J. Mach. Learn. Res.*, 12:2825–2830, 2011.
[11] J. Fan, S. Guo, and N. Hao. Variance estimation using refitted cross-validation in ultrahigh dimensional regression. *J. Roy. Statist. Soc. Ser. B*, 74(1):37–65, 2012.
[12] O. Fercoq, A. Gramfort, and J. Salmon. Mind the duality gap: safer rules for the lasso. In *ICML*, pages 333–342, 2015.
[13] J. Friedman, T. Hastie, H. Höfling, and R. Tibshirani. Pathwise coordinate optimization. *Ann. Appl. Stat.*, 1(2):302–332, 2007.
[14] P. J. Huber. *Robust Statistics*. John Wiley & Sons Inc., 1981.
[15] J. Lederer. Trust, but verify: benefits and pitfalls of least-squares refitting in high dimensions. *arXiv preprint arXiv:1306.0113*, 2013.
[16] X. Li, J. Haupt, R. Arora, H. Liu, M. Hong, and T. Zhao. A first order free lunch for sqrt-lasso. *arXiv preprint arXiv:1605.07950*, 2016.
[17] E. Ndiaye, O. Fercoq, A. Gramfort, V. Leclere, and J. Salmon. Efficient smoothed concomitant Lasso estimation for high dimensional regression. *Arxiv preprint arXiv:1606.02702*, 2016.
[18] E. Ndiaye, O. Fercoq, A. Gramfort, and J. Salmon. Gap safe screening rules for sparse multi-task and multi-class models. In *NIPS*, pages 811–819, 2015.
[19] Y. Nesterov. Smooth minimization of non-smooth functions. *Math. Program.*, 103(1):127–152, 2005.
[20] A. B. Owen. A robust hybrid of lasso and ridge regression. *Contemporary Mathematics*, 443:59–72, 2007.
[21] V. Pham and L. El Ghaoui. Robust sketching for multiple square-root LASSO problems. In *AISTATS*, 2015.
[22] S. Reid, R. Tibshirani, and J. Friedman. A study of error variance estimation in lasso regression. *arXiv preprint arXiv:1311.5274*, 2013.
[23] N. Städler, P. Bühlmann, and S. van de Geer. ℓ_1-penalization for mixture regression models. *TEST*, 19(2):209–256, 2010.
[24] T. Sun and C.-H. Zhang. Comments on: ℓ_1-penalization for mixture regression models. *TEST*, 19(2):270–275, 2010.
[25] T. Sun and C.-H. Zhang. Scaled sparse linear regression. *Biometrika*, 99(4):879–898, 2012.
[26] T. Sun and C.-H. Zhang. Sparse matrix inversion with scaled lasso. *J. Mach. Learn. Res.*, 14:3385–3418, 2013.
[27] R. Tibshirani. Regression shrinkage and selection via the lasso. *J. Roy. Statist. Soc. Ser. B*, 58(1):267–288, 1996.
[28] H. Xu, C. Caramanis, and S. Mannor. Robust regression and lasso. *IEEE Trans. Inf. Theory*, 56(7):3561–3574, 2010.
[29] S. Yun. On the iteration complexity of cyclic coordinate gradient descent methods. *SIAM J. Optim.*, 24(3):1567–1580, 2014.

Robust antenna design through a hybrid inversion strategy combining interval analysis and nature-inspired optimization

M Salucci[1] and T Moriyama[2]

[1] ELEDIA Research Center (ELEDIA@UniTN - University of Trento)
Via Sommarive 9, I-38123 Trento, Italy
[2] ELEDIA Research Center (ELEDIA@UniNAGA - University of Nagasaki)
852-8521, Nagasaki, Japan
E-mail: {marco.salucci, toshifumi.moriyama}@eledia.org

Abstract. The problem of designing robust antenna configurations is cast as the solution of an inverse problem that is addressed by means of a hybrid approach based on Interval Analysis (IA) and Particle Swarm Optimization (PSO). The admissible tolerance error on the surface of reflector antennas or on the element excitations of antenna array is maximized while guaranteeing the upper and lower pattern bounds, including all power patterns potentially generated by the antenna, satisfying user-defined mask constraints. The pattern bounds are analytically computed through the IA for each trial antenna configuration defined by means of the PSO. Representative numerical examples are shown to validate the effectiveness of the proposed approach.

1. Introduction

When designing antennas for some critical applications like advanced communications and radar, there is the need to generate directive radiation patterns with low secondary lobes in order to effectively suppress the undesired signals/interferences and guarantee the system reliability. In the literature, a variety of synthesis techniques have been proposed to define the values of the antenna parameters such to generate the desired radiation pattern in case of both continuous (e.g., reflector antennas) and discrete (e.g., phased array) apertures. However, it has been also verified that the antenna side-lobes are greatly influenced by the accuracy with which one can build the reflector dish and implement the synthesized excitations [1, 2].

To address this problem, robust synthesis techniques have been introduced aimed to achieve the desired performance with a high degree of reliability of the antenna configuration. In this framework, probabilistic approaches considering simplified problem formulations and small errors, approaches considering the power pattern as the superposition of the pattern generated by the nominal antenna plus a "background" power level proportional to the antenna random errors, or Monte Carlo (MC) methods allowing the definition of the maximum tolerance errors guaranteeing the fitting of the desired mask constraints have been proposed [3]-[5].

The limits of the MC analysis, which are the unavoidably high computational burden and consequently the fact that only a subset of all possible error combinations can be contemplated, have been overcome in [6]-[8] for the case of robust phased array design where approaches based on the Interval Analysis [11]-[18] has allowed to analytically compute inclusive power pattern bounds, namely bounds robust to the uncertainties of the array control points. The values of the nominal amplitude coefficients have been optimized by means of the Particle Swarm Optimizer [6] and a deterministic convex minimization procedure [7] for a-priori fixed tolerance errors in order to obtain an interval power pattern lying within pre-imposed mask constraints.

By following the guidelines of [8], this work is aimed to present a method for designing robust beamforming weights of linear antenna arrays as well as, to the first time to the best of the authors' knowledge, reflector antennas where the direct maximization of the tolerance errors on the excitation weights or the dish surface is addressed without any a-priori assumptions on their nominal values or error probability. More in details, assuming the errors be defined as real-valued intervals, the widths of the intervals, namely the admitted tolerance, are maximized while guaranteeing that the bounds of the interval power pattern, computed by means of the IA, satisfy user-defined constraints.

As for the organization of the paper, the problem is mathematically formulated in Sect. II where the integrated PSO-IA-based robust design antenna approach is presented. Representative numerical examples dealing with the design of both discrete array and reflector antennas are shown in Sect. III. Eventually, some conclusions are given in Sect. IV.

2. PSO-based Robust Synthesis Method

Let us consider either the amplitude weights of a phased array or the surface of the dish of a reflector antenna characterized by tolerance errors such that their values can be represented through intervals

$$[\xi_n] = [\xi_n^{\inf};\, \xi_n^{\sup}] \tag{1}$$

where $\xi_n^{\inf}$ and $\xi_n^{\sup}$ are the corresponding left and right interval end-points.

The expression of the field radiated by the antenna turns out to be a complex interval function

$$[E(\theta,\phi)] = \Im\{[\xi_n], n = 0,\ldots,N-1\} \tag{2}$$

defined in terms of the antenna interval parameters which is represented by means of a complex interval value for each angular direction (θ,ϕ) and computed according to [11] or [13] in case of discrete or continuous antenna aperture, respectively.

The interval power pattern is computed as

$$[P(\theta,\phi)] = [E(\theta,\phi)][E(\theta,\phi)]^* \tag{3}$$

and the analytical expressions for the efficient evaluation of its lower $P^{\inf}(\theta,\phi)$ and upper $P^{\sup}(\theta,\phi)$ bounds are given in [11, 13]. In order to design robust antenna configurations, the end-points of the antenna interval parameters $\xi_n^{\inf}$ and $\xi_n^{\sup}$, $n = 0,\ldots,N-1$ are optimized by means of the PSO in order to minimize the following cost function

$$\Phi([\xi_n]) = \Phi^{\sup}([\xi_n]) + \Phi^{\inf}([\xi_n]) + \Phi^{tol}([\xi_n]) \tag{4}$$

where

$$\Phi^{\sup}([\xi_n]) = H^{\sup}(\theta,\phi)\int_0^{2\pi}\int_0^{\pi/2} P^{\sup}(\theta,\phi) - U(\theta,\phi)\, d\theta d\phi \tag{5a}$$

$$\Phi^{\inf}([\xi_n]) = H^{\inf}(\theta,\phi)\int_0^{2\pi}\int_0^{\pi/2} L(\theta,\phi) - P^{\inf}(\theta,\phi)\, d\theta d\phi \tag{5b}$$

being $L(\theta,\phi)$ and $U(\theta,\phi)$ the lower and upper user-defined power pattern constraints and $H^{\sup}(\theta,\phi)=1$ ($H^{\inf}(\theta,\phi)=1$) if $P^{\sup}(\theta,\phi) > U(\theta,\phi)$ ($P^{\inf}(\theta,\phi) < U(\theta,\phi)$) and $H^{\sup}(\theta,\phi)=0$ ($H^{\inf}(\theta,\phi)=0$) otherwise. The term $\Phi^{tol}([\xi_n])$ of (4) is defined such to maximize the robustness of the antenna configuration, expressed as maximum admissible tolerance on the amplitude weights in case of phased arrays or maximum surface distortion in case of reflector antennas.

IOP Conf. Series: Journal of Physics: Conf. Series **904** (2017) 012007 doi:10.1088/1742-6596/904/1/012007

3. Numerical Results

In order to validate the proposed PSO-IA-based robust design method, two representative numerical results dealing with the design of a phased array and a reflector antenna are shown in the following. As for the PSO, the following setting of the control parameters has been taken into account in all simulations [19]: dimension of the swarm equal to 20, inertial weight set to 0.4, and cognitive and social acceleration coefficients to 2.0.

3.1. Phased Antenna Array

Let us consider a linear array with 10 elements end inter-element distance equal to half-wavelength. The mask constraints on the power pattern are shown in Figure 1(a) where $u = \sin\theta$. The values of the interval amplitude configuration of the element weights obtained for the best solution at the end of the PSO-based optimization are shown in Figure 1(b).

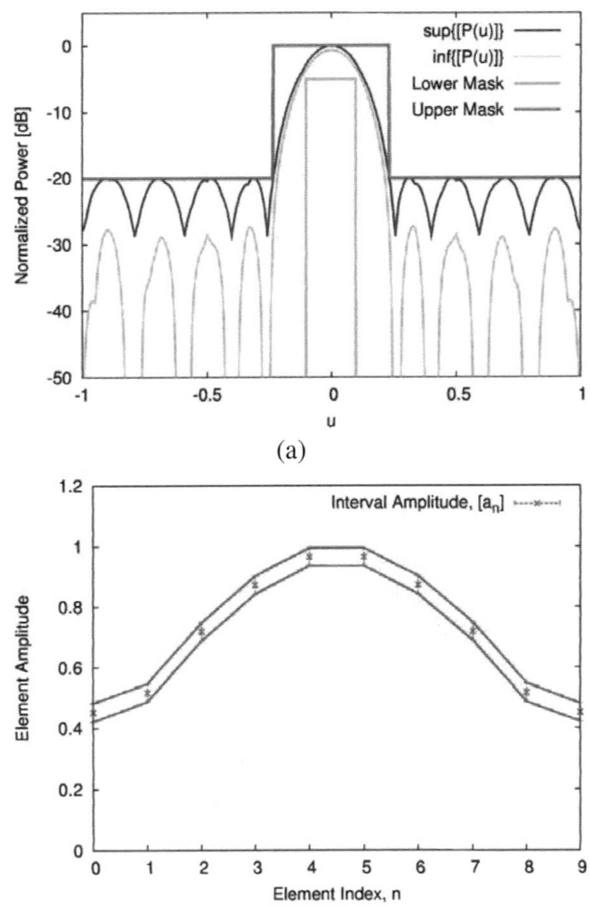

(b)

Figure 1. Plot of (a) the optimized interval amplitude coefficients and (b) the upper and lower bounds of the interval power pattern and of the mask constraints.

As it can be observed from Fig 1(a), the interval power pattern fits the user-defined constraints while the tolerance of the amplitude coefficients have been maximized through the optimization of (4). The simulation converged after K = 100 iterations with a total computational time of 135 [sec] by using a standard CPU (2.4 GHz PC with 2 GB of RAM) and a non-optimized source code.

3.2. Reflector Antennas

In case of continuous aperture antennas, a circular parabolic reflector with diameter equal to 100λ and focal distance 70λ is considered. Like in [13], the far-field pattern is computed by discretizing the aperture into 50 annular sectors and the number of cells on each ring has been determined by imposing a cell area equal to $0.45\lambda^2$.

The power pattern bounds computed by means of the proposed PSO-IA-based method together with the user-defined mask constraints are shown in Figure 2. Such a solution has been obtained through a maximization of the admissible surface error which is characterized by an equivalent root-mean-square error equal to 0.03λ for the best solution of the PSO at convergence. The simulation required K = 500 iterations with a total computational time of 734 [sec].

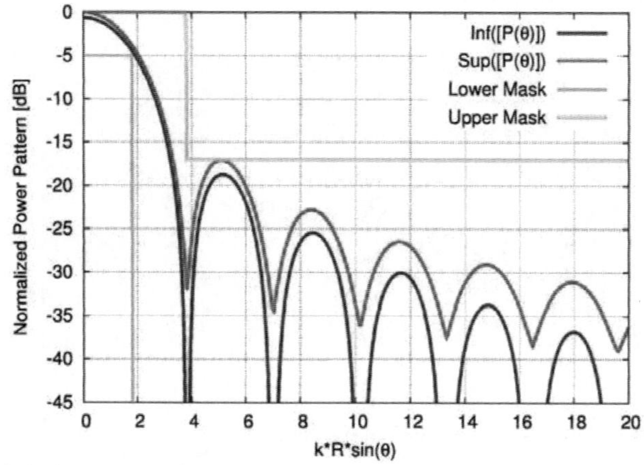

Figure 2. Plot of the upper and lower bounds of the interval power pattern and of the mask constraints.

4. Conclusions

The design of robust antenna configurations generating a reliable power pattern regardless the uncertainties on the antenna amplitude weights in case of phased arrays or on the dish surface in case of reflector antennas has been addressed in this work. The tolerances have been maximized without assuming a-priori error distributions while forcing the bounds of the arising power pattern to satisfy user-defined mask-power constraints. A hybrid approach integrating the PSO for optimizing the end-points of the real-valued interval antenna parameters and the IA-based analysis for analytically defining the interval bounds containing the possibly arising power patterns has been presented and assessed.

References

[1] Ruze J 1952 The effect of aperture errors on the antenna radiation pattern *Nuovo Cimento* (Suppl.) **9** 364-380

[2] Elliott R E 1958 Mechanical and electrical tolerances for two-dimensional scanning antenna arrays *Trans. IRE* **6** 114-120

[3] Ashmead D 1952 Optimum design of linear arrays in the presence of random errors *IRE Trans. Antennas Propag* **4** 81-92

[4] Richards W F and Lo Y T 1975 Antenna pattern synthesis based on optimization in probabilistic sense *IEEE Trans. Antennas Propag* **23** 165-172

[5] Lee J Lee Y Kim H 2005 Decision of error tolerance in array element by the Monte Carlo method *IEEE Trans. Antennas Propag* **53** 1325-1331

[6] Manica L Anselmi N Rocca P Massa A 2013 Robust mask-constrained linear array synthesis through an interval-based Particle Swarm Optimization *IET Microw. Antennas Propag* **7** 976-984
[7] Rocca P Anselmi N Massa A 2014 Optimal synthesis of robust beamformer weights exploiting interval analysis and convex optimization *IEEE Trans. Antennas Propag* **62** 3603-3612
[8] Anselmi N Rocca P Salucci M Massa A 2016 Optimization of excitation tolerances for robust beamforming in linear arrays *IET Microw. Antennas Propag* **10** 208-214
[9] Boche R. 1966 Complex interval arithmetic with some applications *Tech. Report Lockheed Missiles & Space Company, Sunnyvale, California* 1 - 33
[10] Moore R E Kearfott R B Cloud M J 2009 *Introduction to Interval Analysis* SIAM
[11] Anselmi N Manica L Rocca P Massa A 2013 Tolerance analysis of antenna arrays through interval arithmetic *IEEE Trans. Antennas Propag* **61** 5496-5507
[12] Rocca P Manica L Anselmi N Massa A 2013 Analysis of the pattern tolerances in linear arrays with arbitrary amplitude errors *IEEE Antennas Wireless Propag. Lett* **12** 639- 642
[13] Rocca P Manica L Massa A 2014 Interval-based analysis of pattern distortions in reflector antennas with bump-like surface deformations *IET Microw. Antennas Propag* **8** 1277-1285
[14] Rocca P Anselmi N Massa A 2014 Interval arithmetic for pattern tolerance analysis of parabolic reflectors *IEEE Trans. Antennas Propag* **62** 4952-4960
[15] Anselmi N Salucci M Rocca P Massa A 2016 Generalized interval-based analysis tool for pattern distortions in reflector antennas with bump-like surface deformations *IET Microw. Antennas Propag* **10** 909-916
[16] Rocca P Poli L Anselmi N Salucci M Massa A 2016 Predicting antenna pattern degradations in microstrip reflectarrays through interval arithmetic *IET Microw. Antennas Propag* **10** 817-826
[17] Anselmi N Rocca P Salucci M Massa A 2016 Power pattern sensitivity to calibration errors and mutual coupling in linear arrays through circular interval arithmetics *Sensors* **16** 1-14
[18] Tenuti L Anselmi N Rocca P Salucci M Massa A 2017 Minkowski sum method for planar arrays sensitivity analysis with uncertain-but-bounded excitation tolerances *IEEE Trans. Antennas Propag* **65** 167-177
[19] Rocca P Benedetti M Donelli M Franceschini D Massa A 2009 Evolutionary optimization as applied to inverse scattering problems *Inv. Probl* **25** 1-41

Acknowledgment
This work has been partially supported by the SIRENA project (2014-2017) funded by DIGITEO (France) under the "Call for Chairs 2014".

Brain waves-based index for workload estimation and mental effort engagement recognition

A. Zammouri[1,2,*], **S. Chraa-Mesbahi**[3], **A. Ait Moussa**[1], **S. Zerouali**[4], **M. Sahnoun**[4], **H. Tairi**[3] and **A. M. Mahraz**[3]

[1] Department of Computer Science, Faculty of Sciences, Mohammed First University, Av. Med VI, P.B. 717, 60000 Oujda, Morocco
[2] Centre de Recherche de l'Ecole des Hautes Etudes d'Ingénierie, 60000, Oujda, Maroc ,
[3] University of Sidi Mohammed Ben Abdellah, Department of Informatics, Faculty of Science Dhar-Mahraz, P.B.1796, 30000 Atlas-Fez, Morocco
[4] CESI-Centre Nord-Ouest, 1 Rue Marconi – CS 30285, 76137 Mont Saint Aignan, France

Abstract. The advent of the communication systems and considering the complexity that some impose in their use, it is necessary to incorporate and equip these systems with a certain intelligence which takes into account the cognitive and mental capacities of the human operator. In this work, we address the issue of estimating the mental effort of an operator according to the cognitive tasks difficulty levels. Based on the Electroencephalogram (EEG) measurements, the proposed approach analyzes the user's brain activity from different brain regions while performing cognitive tasks with several levels of difficulty. At a first time, we propose a variances comparison-based classifier (VCC) that makes use of the Power Spectral Density (PSD) of the EEG signal. The aim of using such a classifier is to highlight the brain regions that enter into interaction according to the cognitive task difficulty. In a second time, we present and describe a new EEG-based index for the estimation of mental efforts. The designed index is based on information recorded from two EEG channels. Results from the VCC demonstrate that powers of the Theta [4-7 Hz] (θ) and Alpha [8-12 Hz] (α) oscillations decrease while increasing the cognitive task difficulty. These decreases are mainly located in parietal and temporal brain regions. Based on the Kappa coefficients, decisions of the introduced index are compared to those obtained from an existing index. This performance assessment method revealed strong agreements. Hence the efficiency of the introduced index.

1. Introduction

The ambition behind communication systems is evolving the communication between human and machine in order to improve his/her everyday life. With the innovative aspect brought by the Artificial Intelligence (AI), communication systems have increasingly integrated the human's everyday life. In the context of rehabilitation and health care, these systems represent innovative tools for the assistance of people with disabilities or an extreme dependence. However, designing systems communicating with humans must incorporate dynamic and adaptive characters. In the context of assistive technologies, a crucial aspect in designing such systems is to take into account the specificities of each user and to propose solutions adapted to his/her mental and motor abilities [1]. In order to measure affective status of an individual, several approaches have been proposed in literature; the more robust are that based on biological signals [2],[3]. Amongst these methods we distinguish Heart Rate Variation (HRV), Galvanic Skin Response (GSR) and the brain electrical activity [4]. However,

IOP Conf. Series: Journal of Physics: Conf. Series **904** (2017) 012008 doi:10.1088/1742-6596/904/1/012008

amongst these approaches, only analysing the electrical brain activity has the potential to precisely reflect the affective status of an individual. For about two decades, studying and analysing the human's brain activity have been closely related to the concept of Brain-Computer Interface (BCI). In its definition, a BCI represents a direct communication and control system between the human and any electronic device [5]. BCI provides a means of interacting and acting with the physical surrounding without using the standard neuromuscular channels [1],[6]. The user's intentions are mediated through brain signals in place of nerves and muscles. This type of peripheral, which is completely different from the usual Human-Machine Interfaces (HMI), can be designed to assist, improve or repair human cognition functions [7],[8].

In the context of cerebral interfaces, especially those based on EEG signals, features allowing distinguishing mental tasks and mental states vary from the amplitude of the brain signal, the Power Spectral Density (PSD) and time-frequency features. Approaches using these features take advantage of the spectral aspect of the EEG signal. Indeed, through a transform into the frequency domain, the EEG signal can be decomposed in different ranges of frequencies namely, Delta [0.5-3 Hz] (δ), Theta [4-7 Hz] (θ), Alpha [8-12 Hz] (α), Beta [13-30 Hz] (β) and Gamma [>30 Hz] (γ) [9].

In this work, we aim at developing a passive BCI to estimate the cognitive load level and the mental effort engagement of an individual. Our experimental protocol aims at measuring EEG signals on four subjects while answering matrices products with different levels of difficulties. At a first time, the proposed approach combines the power spectral density analysis of EEG signals and a variances comparison-based classifier (VCC) in order to illustrate changes in the frequency features according to the cognitive task difficulties. In a second time, we introduce a new indicator of mental effort engagement which reacts sensitively to changes of the cognitive task difficulty levels. Our approach is based on both the θ and α frequency components. This paper is organized as follows. In the second section we present a background of related works from literature. In the third section we describe our approach for estimating the cognitive workload and mental effort engagement. Experimental results are detailed and discussed in the fourth section. Finally conclusions are drown in the fifth section.

2. Background and related works

In psychology, the use of EEG measures represents the basis background of a large number of researches in the cognitive workload analysis. However, there is still a doubt on the effect of cognitive workload on electroencephalographic signals. In works of Smith and Gevins [10], an approach is presented to assess EEG recordings of subjects performing n-back tests and other computer interaction tasks. This approach is based on the extraction of spectral characteristics of θ and α frequency bands in four seconds time-windows, and then, neural networks are applied for the discrimination of three different levels of cognitive workload. In another work related to the cognitive workload assessment, Putze [11] proposed a multimodal approach to estimate the different levels of the cognitive workload. Authors propose, in addition to EEG data, measuring the skin conductance and breathing. These measures are then classified on time-windows of 60 seconds. This approach was applied in the context of assessing the cognitive workload in a driving simulator. In the realm of using EEG signals to estimate mental states, Pal et al., have introduced an unsupervised approach, based on the Mahalanobis distance, to recognize departs in drowsiness of drivers in an environment based on Virtual Reality (VR) [12]. This approach generates, in each driving session, a statistical model of the driver's alertness based on θ and α frequency bands from the occipital brain area. Through this work, authors demonstrated that deviating from de alert model varies according to the driving performance.

Estimating the cognitive workload from EEG requires objective methods to determine the cognitive overload of the brain mental fatigue during the work performance. In [13], authors have determined an index ρ, based on the frontal (F_Z) θ oscillation and the parietal (P_Z) α oscillation:

- $$\rho = \frac{\theta(F_Z)}{\alpha(P_Z)} \quad (1)$$

The ρ index increases with the number of tasks to perform simultaneously and time awake. The main results of this approach are focused on the information from EEG spectrum and particularly the θ oscillation increasing in the frontal area and the α oscillation decreasing in the parietal area.

When performing certain cognitive tasks, some changes are observed in EEG signals. We talk about Event Related Synchronization (ERS) and Desynchronization (ERD). According to Antonenko et al., [14] and Fink et al., [15] ERS and ERD rates measure the oscillatory dynamics. According to the Antonenko, the ERD/ERS ratio reflects the decrease (ERD) or the increase (ERS) of the α power during a time window which is compared to a reference time window. The ratio r = ERD/ERS is defined as follows:

- $$r = \frac{P_b^{\alpha} - P^{\alpha}}{P_b^{\alpha}} * 100 \quad (2)$$

where P^{α} represents the α band power during a given time window. P_b^{α} is the α band power during the reference time window. According to Antonenko, a positive r ratio reflects an increase in the α power, i.e. ERS. According to Grabner et al., [16], performing a cognitive task of a high level of difficulty increases the ERD in the α band. This implies a decrease in the power spectrum.

3. Adopted model

3.1. Protocol and experimentations

To induce different levels of cognitive workload we used as cognitive task solving matrices products with different difficulty levels. The experiments conducted in this study consist of two sessions. The cognitive task difficulty varies in an ascending way from the first session to the second one. Arising the difficulty level lies in increasing sizes of matrices.

EEG data used in our study were recorded on students of the Department of Mathematics and Computer Science at the Mohammed First University, Morocco. The population consists of four male subjects right handed and aged between 23 and 33years old. EEG data were measured in clinical conditions at the Neurology Department, Mohammed Sixth Hospital Center, Morocco. EEG measures were recorded based on the EBNeuro acquisition system with 7 electrodes at sampling rate of 128 Hz. The EEG data were filtered using the band-pass [1-30Hz].

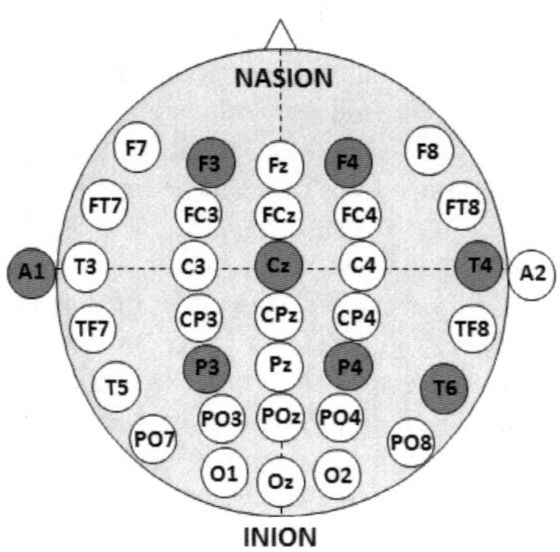

Fig. 1 Placement of the used electrodes according to the extended 10-20 international system.

3.2. Cognitive workload estimation approach

The cognitive workload is characterized particularly by evolutions of brain activities in the frequency range of [1-30Hz]. Therefore, we suggest focusing on the evolution of θ and α brain waves in the central, parietal and temporal areas. At a first time, we aim at eliminating artifacts using the Blind Source Separation (BSS) method applied to the 7 used electrodes. Meaning the Short Time Fourier Transform (STFT), the resulting EEG signal's power spectrum is computed based on the Welch periodogram method [17]. The signal is segmented into several overlapping and equal parts. Therefore the STFT is computed on each segment. The result is the arithmetic average of the segments transform. This method is applied every 10 seconds and on each EEG channel with a moving window fixed at 2s. This window duration seems appropriate in the frequency range of [1-30 Hz] for the STFT computation. The use of the BBS method at the beginning of our approach aims to reduce the effect of artifacts on the calculation of the spectral power on the α band. To estimate the workload difference (changes of θ and α band powers) between the two sessions, a comparison is made using a VCC applied to each considered band power.

Let $P^*_{m_1}$and $P^*_{m_2}$be two independent samples of the power spectral density of each of θ and α (represented by $*$) on a given EEG channel. We denote by $\overline{P^*_{m_1}}$ and $\overline{P^*_{m_2}}$ the means of $P^*_{m_1}$and $P^*_{m_2}$ respectively. Sizes of samples are respectively $n^*_{m_1}$and $n^*_{m_2}$. We denote by ${S^2}^*_{m_1}$and ${S^2}^*_{m_2}$ the experimental variances of these samples respectively. Through the VCC we aim to determine whether the two samples belong to the same population. Therefore, two hypotheses are defined. The Null Hypotheses (H^*_0) corresponds to ${S^2}^*_{m_1} = {S^2}^*_{m_2}$; while the Alternative Hypothesis corresponds to the case of ${S^2}^*_{m_1} \neq {S^2}^*_{m_2}$. Hence the statistic of the VCC is defined as follows:

- $$\mathrm{F}^* = \frac{{\mathrm{S}^2}^*_{\mathrm{m}_1}}{{\mathrm{S}^2}^*_{\mathrm{m}_2}} \qquad (3)$$

The defined statistic is observed along a Fisher law with $\nu = n^*_{m_1} + n^*_{m_2} - 2$ degrees of freedom. The comparison of variances can be tested based on a Confidence Interval ($CI_{\rho,\nu}$). If $F^* \in CI_{\rho,\nu}$ we can reject H^*_0 with a ρ risk of having reject it. If $F^* \notin CI_{\rho,\nu}$ we accept H^*_0. The theoretical values of the $CI_{\rho,\nu}$ boundaries are extracted from the Fisher-Snedecor table according to values of ρ and ν.

Using this classification method allows to determine whether the mental efforts produced during the two sessions of the experimental protocol are of the same class or not. Moreover, it allows to determine brain regions in which cognitive workload changes mainly took place. In order to better estimate and assess the mental efforts of each participant, we introduce the new z index which is based on the parietal θ and the frontal α oscillations:

- $$\mathrm{z} = \frac{\theta(\mathrm{P4})}{\alpha(\mathrm{F4})} \qquad (4)$$

4. Results and discussion

Results presented in Fig. 2 demonstrate the sensitivity of the **z** index in terms of estimating the mental efforts and the cognitive workload expended to perform the cognitive tasks of the used experimental protocol. These results show, for all participants, a significant decrease of the **z** index value during the second session, i.e. when increasing the cognitive task difficulty. Through a representative example (subject 2), presented in Fig. 3, we illustrate the **z** index behavior while performing the first session. Indeed, during the first time window, i.e. the first 10 seconds, the **z** index revealed a large value. During this time window, the subject does not perform any arithmetic operation; he read the session statement. Hence, no high level of vigilance is required during this time window. During the second time window, the user started resolving matrices products. This transition from reading the session statement to the performance of the first computations has strongly decreased the **z** index.

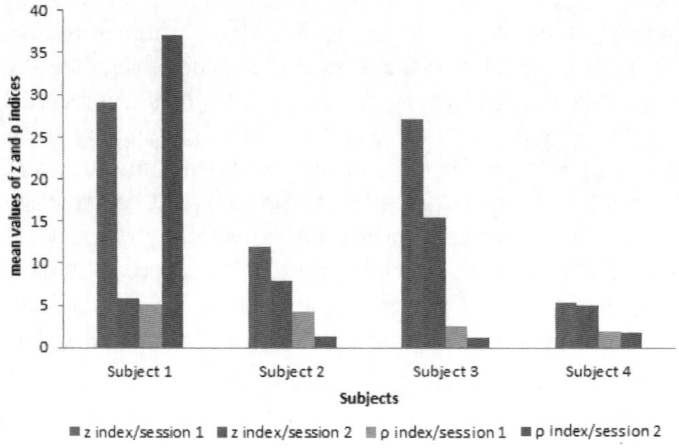

Fig. 2 Values (averages) of the *z* index obtained for all subjects during the two sessions of the experimental protocol.

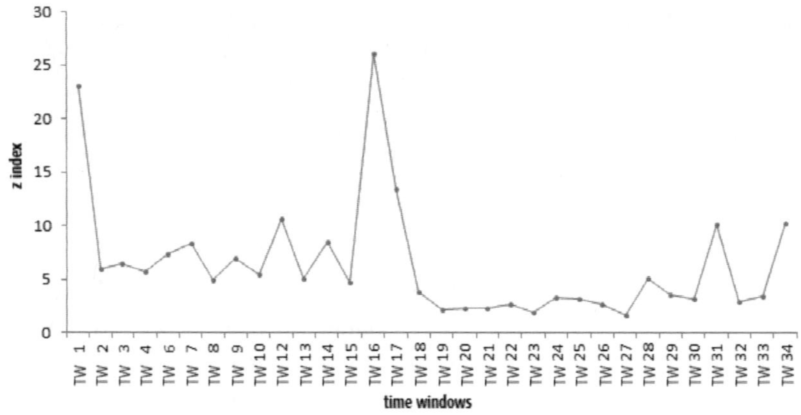

Fig. 3 Behavior of *z* index obtained from a representative example (subject 2) during the first sessions

From the second time window to the fifteenth one, the subject performed matrices products with a low level of difficulty. During time windows 16 and 17 the subject wrote his responses on the answers sheet. These two time windows are characterized by large values of the *z* index. Starting from the time window 18, the subject performed matrices products with a significant high level of difficulty. One can note that the *z* index values during these time epochs are more decreased than those when the subject performed cognitive task with a low level of difficulty. Finally, during the last four time windows the subject reported his responses on the answers sheet. This is reflected by values significantly large. Hence the sensitivity of the *z* index to describe changes in the cognitive workload levels.

In order to evaluate our new index, a comparison to the ρ index, introduced by Holm [13], is made. Results from Fig. 2 present, in averages, values of the two indices, z and ρ, during the two sessions. These results show similar behaviors of the two indices except the case of subject 1 for whom the ρ index increased during the second session. Unlike the Holm's assumption, results from Fig. 3 show a decrease in the ρ index. This implies that the α wave decreases in the parietal brain area when increasing the difficulty of the cognitive task. Our assumption is illustrated by the obtained results presented in Fig. 2. To better assess the performance of the introduced index, we used the Kappa coefficient [18] to compute the agreement rate between decisions obtained from our index and those obtained from the r index. The use of the Kappa coefficient identifies, beyond chance, the quality of a classifier. Evaluation decisions are made by computing the coefficient presented in Eq. 5. $P_r(a)$ represents the degree of agreement between decisions of the two indices. $P_e(a)$ represents the probability of a random agreement between the compared indices. Values of the Kappa coefficient range from -1 to 1. The higher the value, the stronger the agreement.

- $k = \frac{P_r(a)-P_r(e)}{1-P_r(e)}$ (5)

Result of this assessment process revealed a value k = 0.58 which reflects a strong agreement between our index and the ERD/ERS ratio.

5. Conclusion

In this paper we have presented an EEG-based index to estimate brain workload and mental effort engagement. This indicator is based on spectral density using the Welch periodogram. EEG powers were calculated in the θ and α bands in even central, parietal and temporal brain areas. Results obtained on four subjects have shown that increasing the cognitive task's difficulty reduces the power spectral density of θ and α bands in the specified areas. In a future extension of this work, a large data set should be considered in order to accurately validate the obtained results. Moreover, we plan to determine thresholds on θ and α bands which can be used to label the subject's cognitive states.

6. References

[1] Kalunga, E.K., Chevallier, S., Rabreau, O., Monacelli, E., Hybrid interface: Integrating BCI in multimodal human-machine interfaces, *in*: Proceedings of the IEEE Conf. on Advanced Intelligent Mechatronics (AIM), 2014.

[2] Healey, J., Picard, R. W., 2005. Detecting Stress During Real-World Driving Tasks Using Physiological Sensors, IEEE Trans. Intell. Transp. Syst. 6, 156-166.

[3] Wobrock, D., Frey, J., Graeff, D., de la Rivière, J.-B., Castet, J. Lotte, F., Continuous Mental Effort Evaluation During 3D Object Manipulation Tasks Based on Brain and Physiological Signals, *in*: Proceedings of the Conf. on Human-Computer Interaction (INTERACT) 2015.

[4] Mühl, C., Jeunet, C., Lotte, F., 2014. EEG-based workload estimation across affective contexts. Front. Neurosci. 8.

[5] Wolpaw, J. R., Birbaumer, N., McFarland, D. J., Pfurtscheller, G., Vaughan, T. M., 2002. Brain-computer interfaces for communication and control, Clin. Neurophysiol. 113, 767–791.

[6] Pfurtscheller, G., Flotzinger, D., Kalcher, J., 1993. Brain-Computer Interface—a new communication device for handicapped persons. J. Microcomput. Appl. 16, 293–299.

[7] Leeb, R., Perdikis, S., Tonin, L., Biasiucci, A., Tavella, M., Creatura, M., Molina, A., Al-Khodairy, A., Carlson, T., Millán, J. d.R., 2013. Transferring brain–computer interfaces beyond the laboratory: Successful application control for motor-disabled users. Artif. Intell. Med. 59, 121–132.

[8] Pires, G., Nunes, U., Castelo-Branco, M., 2012. Evaluation of Brain-computer Interfaces in Accessing Computer and other Devices by People with Severe Motor Impairments, Proced. Comput. Sci. 14, 283–292.

[9] Başar, E., Başar-Eroglu, C., Karakaş, S., Schürmann, M., 2001. Gamma, alpha, delta, and theta oscillations govern cognitive processes. Int. J. Psychophysiol. 39, 241–248.

[10] Gevins, A., Smith, M. E., 2003. Neurophysiological measures of cognitive workload during human-computer interaction, Theor. Issues Ergon. Sci. 4, 113–131.

[11] Putze, F., Jarvis, J.-P., Schultz, T., Multimodal Recognition of Cognitive Workload for Multitasking in the Car, *in*: Proceedings of the IEEE Conf. on Pattern Recognition (ICPR) 2010.

[12] Pal, N.R., Chuang, C.-Y., Ko, L.-W., Chao, C.-F., Jung, T.-P., Liang, S.-F., Lin, C.-T., 2008. EEG-Based Subject- and Session-independent Drowsiness Detection: An Unsupervised Approach. EURASIP J. Adv. Signal Process.

[13] Holm, A., Lukander, K., Korpela, J., Sallinen, M., Müller, K.M.I., 2009. Estimating Brain Load from the EEG. Sci. World J. 9, 639–651.

[14] Antonenko, P., Paas, F., Grabner, R., van Gog, T., 2010. Using Electroencephalography to Measure Cognitive Load. Educ. Psychol. Rev. 22, 425–438.

[15] Fink, A., Grabner, R.H., Neuper, C., Neubauer, A.C., 2005. EEG alpha band dissociation with increasing task demands. Cogn. Brain Res. 24, 252–259.

[16] Grabner, R.., Fink, A., Stipacek, A., Neuper, C., Neubauer, A.., 2004. Intelligence and working memory systems: evidence of neural efficiency in alpha band ERD. Cogn. Brain Res. 20, 212–225.

[17] Welch, P., 1967. The use of fast Fourier transform for the estimation of power spectra: A method based on time averaging over short, modified periodograms. IEEE Trans. Audio Electroacoustics. 15, 70–73.

[18] Cohen, J.A., 1960. A Coefficient of Agreement for Nominal Scales. Educ. Psychol. Meas. 20, 37–46

IOP Conf. Series: Journal of Physics: Conf. Series **904** (2017) 012009 doi:10.1088/1742-6596/904/1/012009

A fast subgradient algorithm in image super-resolution

D. Lazzaro[1], E. Loli Piccolomini[1], V. Ruggiero[2], F. Zama[1]

[1] Department of Mathematics, University of Bologna

[2] Department of Mathematics and Computer Science, University of Ferrara

E-mail: `fabiana.zama@unibo.it`

Abstract. In this paper we propose an ϵ−subgradient method for solving a constrained minimization problem arising in super-resolution imaging applications. The method, compared to the state-of-the-art methods for single image super-resolution on some test problems, proves to be very efficient, both for the reconstruction quality and the computational time.

1. Introduction

Image super-resolution reconstruction is the process of obtaining High Resolution (HR) images from observed Low Resolution (LR) images. The problem of super-resolution is of great importance in many applications, such as medicine or object recognition (face, bar codes, ...), where the electronic imaging devices are equipped with low resolution cameras, while HR images are finally desired. The super-resolution process can be performed from a single image (SISR) or from multiple images of the same scene (MISR), such as in the case of videos or medical imaging devices for example. In this paper, we will consider only the case of SISR.

The SIRS problem is ill-posed, since identical LR images can be generated from different HR images. Furthermore, in addition to being connected by a down-sampling operator, the HR and LR images are related by a Point Spread Function (as a simple convolution, for example), that is an ill-posed operator. The different approaches, present in literature for image super-resolution, can generally be grouped into four categories: interpolation-based algorithms, example-based algorithms, sparse-representation-based algorithms and reconstruction-based algorithms (see [5] and the references therein). The reconstruction-based algorithms are aimed at solving a minimization problem, where the objective function is the sum of a fidelity term related to the data noise and a regularization term, representing the prior on the solution and, at the same time, enabling to face the ill-posedness of the problem. This formulation has the advantage of simultaneously exploiting the degradation model and the information contained in the prior, thus reducing noise and artifacts in the HR reconstructed image [5, 12].

We consider a reconstruction-based algorithm to address a minimization problem where the fidelity term is the ℓ_2-norm, the regularization term is the discrete Total Variation (TV) function and the solution is constrained to be nonnegative. This model has been widely treated for the deblurring applications and many algorithms have been proposed for its solution, such as, for example, [11, 13, 14]. The Bregman method proposed in [14] has been later used in [12] for image super-resolution. In this paper we propose a scaling ϵ-subgradient method presented in [2] for deblurring of images corrupted by Poisson noise. The method has a fast computational

IOP Conf. Series: Journal of Physics: Conf. Series **904** (2017) 012009 doi:10.1088/1742-6596/904/1/012009

cost, since it doesn't require any linear system solution. We test the algorithm on two problems related to synthetic and natural images and compare the results with some state-of-the-art methods for super-resolution.

The paper is organized as follows. In section 2 we present the mathematical problem formulation; in section 3 we describe the algorithm and finally in section 4 we report the results of some numerical experiments and the final conclusions.

2. Numerical model and notation

Let $G \in \mathbb{R}^{M\times N}$ be a low resolution observed image and let $X \in \mathbb{R}^{(s\cdot M)\times(s\cdot N)}$ be the high resolution image to be recovered, with up-sampling factor $s > 1$. The mathematical model of a LR image formation can be written as:

$$g = SHx + \eta \tag{1}$$

where $g \in \mathbb{R}^m$ and $x \in \mathbb{R}^n$ are obtained after vector reordering of the matrices G and X respectively, hence $m = M \cdot N$ and $n = s^2 \cdot M \cdot N$. The matrix H represents the discrete blurring operator acting on the HR data x and S is the down-sampling operator mapping HR blurred data into the LR data. The vector $\eta \in \mathbb{R}^m$ represents additive data noise.

The numerical model proposed in the reconstruction-based super-resolution algorithms is given by the following constrained minimization problem:

$$min_{x\in\mathbb{R}^n} \|SHx - g\|_2^2 + \beta TV(x) \quad s.t.\ x \geq 0 \tag{2}$$

where the nonnegativity constraint on the solution is a physical requirement on the components of the image to be recovered. Here $\beta > 0$ is the regularization parameter and the discrete TV can be written as

$$TV(x) = \sum_{i=1}^{n} \|A_i x\|$$

where $A_i \in R^{2\times n}$ is the forward difference approximation of the gradient of x at the pixel i. Under the usual assumption on the blurring operator, such as

$$H_{i,j} \geq 0 \ ; \ \sum_{i=1}^{n} H_{i,j} > 0 \ , \ \forall j = 1, ..., n \ ; \ \sum_{j=1}^{n} H_{i,j} > 0 \ , \ \forall i = 1, ..., n, \tag{3}$$

the problem (2) is coercive and then a solution exists. Let's denote $\mathbf{1}$ as the vector with all entries equal to 1. Since $H\mathbf{1} \neq \mathbf{0}$ and $\mathcal{N}(SH) = \{x : (Hx)_i = 0 \text{ for } i \in I\}$, where $I \neq \emptyset$ is the set of the non zero columns of S, the intersection of the space of constant images $\alpha\mathbf{1}$ (which is the null space of $TV(x)$) with the null space of SH is just the zero vector.
In the following, we denote by $A \in \mathbb{R}^{2n\times n}$ the block matrix $(A_1^T, A_2^T, ...A_n^T)^T$. Furthermore, given a symmetric positive definite matrix D, $P_{D,C}(x)$ denotes the projection on the closed set C with respect to the metric induced by D, defined as $P_{D,C}(x) = \text{argmin}_{z\in C} \|z - x\|_D^2$, where $\|y\|_D^2 = y^T Dy$. For $D = I$, we have the standard Euclidean projection on the set C. The standard Euclidean norm is denoted by $\|\cdot\|$.

3. The scaled subgradient algorithm

The considered application is a special case of the problem

$$\min_{x\in\mathbb{R}^n} f_0(x) + f_1(Ax) + \Phi(x), \tag{4}$$

where $f_0(x) = \frac{1}{2}\|SHx - g\|^2$ is the discrepancy function, $f_1(Ax) = \beta TV(x)$ is the discrete Total Variation and $\Phi(x)$ is the indicator function $\iota_X(x)$ of the non-negative orthant X. Using the Fenchel transform of $f_1(Ax)$, which is the indicator function $i_Y(y)$ of the bounded and convex set $Y = \{(y_1^T, ...y_n^T)^T | y_i \in \mathbb{R}^2, \|y_i\| \leq 1\}$, from the equality $f_1(Ax) = \beta \max_{y\in Y}(Ax)^T y$, we obtain the equivalent primal–dual formulation of the problem (2) and we can address the numerical solution by the class of Primal–Dual Hybrid Gradient (PDHG) methods (see [17, 4] and references therein). Recently in [2] a variable metric has been introduced in a PDHG method and, by interpreting this scheme as an ϵ-subgradient iteration, a dynamic computation for the primal stepsize has been proposed. For the problem (2), the basic iteration of the Scaled Primal–Dual Hybrid Gradient (SPDHG) method requires only matrix–vector operations and simple projections:

$$y^{(k+1)} = P_{I,Y}(y^{(k)} + \tau_k \beta A x^{(k)}) \tag{5}$$

$$u^{(k)} = d^{(k)} + \beta A^T y^{(k+1)} \tag{6}$$

$$x^{(k+1)} = P_{D_k^{-1},X}\left(x^{(k)} - \frac{\alpha_k}{\min(1, \|u^{(k)}\|_{D_k})} D_k u^{(k)}\right) \tag{7}$$

where $d^{(k)} = H^T S^T (SHx^{(k)} - g)$, $\{D_k\}$ is a sequence of symmetric positive definite matrices with bounded eigenvalues and $\{\tau_k\}$, $\{\alpha_k\}$ are the dual and primal steplength sequences respectively. Method (5)–(7) is a special case of a scaled forward-backward ϵ-subgradient method, where the forward step (5)–(6) provides an ϵ-subgradient of $f_0 + f_1 \circ A$ and the backward or proximal step (7) of $i_X(x)$ is the projection on the non-negative orthant with respect to the variable metric induced by D_k^{-1}. The key point of the interpretation of the forward step is based on the additivity of the ϵ-subgradient and on the fact that the vector $y^{(k+1)}$ defined in (5) belongs to the domain of the Fenchel transform of f_1. Consequently, $\beta A^T y^{(k+1)}$ is an ϵ_k-subgradient of $f_1 \circ A$ at $x^{(k)}$ with $\epsilon_k = f_1(Ax^{(k)}) + \iota_Y(y^{(k+1)}) - \beta {y^{(k+1)}}^T Ax^{(k)}$. Furthermore, $\epsilon_k \leq \frac{\text{diam}(Y)^2}{2\tau_k}$ (see [3, Lemma 1]) and, for an *a-priori* selected divergent sequence $\{\tau_k\}$, $\epsilon_k \to 0$, as $k \to \infty$.
In the backward step we use an adaptive stepsize selection strategy for the sequence $\{\alpha_k\}$, inspired by the idea of *level algorithm* of [9]. The resulting algorithm, detailed in Algorithm 1, is the version of SSL algorithm described in [2] for the problem (2). In particular, in the scheme we have $f_k^{rec} = \min_{i=0,...,k}(f_0(x^{(i)}) + f_1(Ax^{(i)}))$, while l is the number of times that the value f^{lev} has been updated and $k(l)$ is the iteration where the l-th updating occurred. Finally, σ_k is the cumulative path length between two successive updates of f^{lev}. Steps 2-5 aim to provide in f_k^{lev} an estimate of the optimal function value at the iterate k, which is used as target level for the successive iterates until the objective function value is sufficiently close to it or the iterates move through a long path without approaching it. In the first case, i.e. when the inequality at Step 3 is satisfied, f_k^{lev} is reduced at Step 5 by subtracting the positive quantity δ_l to the best value obtained so far, f^{rec}. In the other case, when the inequality at Step 4 is satisfied, the estimated difference from the optimal value δ_l is reduced and, as a consequence of Step 5, the target level f_k^{lev} is increased. Assuming to choose an a-priori dual stepsize τ_k such that $\lim_{k\to\infty} \tau_k = \infty$ and a sequence of positive definite scaling matrices $\{D_k\}$ converging to the identity matrix as $k \to \infty$, the convergence of the SPDHG method with adaptive updating rule for α_k is assured by Corollary 5.2 in [2]. For completeness, we report the convergence proposition for the problem (2).
Proposition. Let $\{x^{(k)}\}$ be the sequence generated by Algorithm 1, where $u^{(k)}$ in (7) is given by (6) with $d^{(k)} = H^T S^T (SHx^{(k)} - g)$. Assume that that there exists $\rho > 0$ such that $\|d^{(k)}\| \leq \rho$ for all k. Define $L_k = \max(\|D_k\|, \|D_k^{-1}\|)$ and assume that $\lim_{k\to\infty} \tau_k = \infty$, $L_k \leq \sqrt{1 + \gamma_k}$, $\gamma_k = \mathcal{O}\left(\frac{1}{k^q}\right)$ with $q > 1$. Then we have $\liminf_{k\to\infty} f_0(x^{(k)}) + f_1(Ax^{(k)}) = \min_{x\in X}(f_0(x) + f_1(Ax))$.

Proof. Since $d^{(k)} = \nabla f_0(x^{(k)})$, by Lemma 1 in [3] we have $u^{(k)} \in \partial_{\epsilon_k}(f_0 + f_1 \circ A)(x^{(k)})$, where $\epsilon_k = f_1(Ax^{(k)}) + \iota_Y(y^{(k+1)}) - \beta {y^{(k+1)}}^T Ax^{(k)}$, with $\epsilon_k \leq \frac{\text{diam}(Y)^2}{2\tau_k}$. Since $\lim_{k\to\infty} \tau_k = \infty$, we have $\lim_{k\to\infty} \epsilon_k = 0$. Choosing a sequence of null subgradients at $x^{(k)}$ for i_X, by Theorem 4.1 in [2], we obtain the result.

Algorithm 1 Scaled Primal–Dual Hybrid Gradient (SPDHG) method for problem (2)

Choose $B > 0$, $\nu_1, \nu_2 \in (0,1)$, $f_{-1}^{rec} = \infty$; $k = 0$, $l = 0$, $k(l) = 0$, $\delta_0 > 0$; choose $x^{(0)} \in X$.

FOR $k = 0, 1, 2, \ldots$

STEP 0. Computation of ϵ-subgradient $u^{(k)}$ of $f_0(x^{(k)}) + f_1(Ax^{(k)})$ with (5)-(6)

STEP 1. Computation of $f_0(x^{(k)}) + f_1(Ax^{(k)})$

STEP 2. If $f_0(x^{(k)}) + f_1(Ax^{(k)}) < f_{k-1}^{rec}$, then $f_k^{rec} = f_0(x^{(k)}) + f_1(Ax^{(k)})$ else $f_k^{rec} = f_{k-1}^{rec}$.

STEP 3. If $f_0(x^{(k)}) + f_1(Ax^{(k)}) < f_{k(l)}^{rec} - \nu_1\delta_l$, then $k(l+1) = k$, $\sigma_k = 0$, $\delta_{l+1} = \delta_l$, $l = l+1$ and go to Step 5.

STEP 4. If $\sigma_k > B$, then $k(l+1) = k$, $\sigma_k = 0$, $\delta_{l+1} = \nu_2\delta_l$, $l = l+1$.

STEP 5. Set $f_k^{lev} = f_{k(l)}^{rec} - \delta_l$.

STEP 6. Update the stepsize and compute the new iterate

$$\alpha_k = \frac{f_0(x^{(k)}) + f_1(Ax^{(k)}) - f_k^{lev}}{\max(1, \|u^{(k)}\|_{D_k})},$$

$$x^{(k+1)} = P_{D_k^{-1},X}\left(x^{(k)} - \alpha_k D_k \frac{u^{(k)}}{\max(1, \|u^{(k)}\|_{D_k})}\right), \tag{8}$$

STEP 7. $\sigma_{k+1} = \sigma_k + \alpha_k$ and go to Step 1.

END

In order to fully define the SPDHG method, we focus on the strategy to compute a suitable scaling matrix D_k, adapting to our case the split gradient strategy proposed in [1] for nonnegatively constrained differentiable problems, which demonstrated to be very effective in several applications (see [16, 6] and references therein). The key point of this approach consists in finding a subgradient decomposition of the form $u^{(k)} = V(x^{(k)}) - U(x^{(k)})$ with $V(x^{(k)}) > 0$ and $U(x^{(k)}) \geq 0$ for all k and then defining D_k in (7) as a diagonal scaling matrix whose entries are the projection of $x_i^{(k)}/V_i(x^{(k)})$ onto the set $[1/\sqrt{1+\gamma_k}, \sqrt{1+\gamma_k}]$. Thus, we have to find a decomposition of the vector $u^{(k)} = \nabla f_0(x^{(k)}) + \beta A^T y^{(k+1)}$ as the difference of two nonnegative terms and we put $V(x^{(k)}) = H^T S^T S H x^{(k)} + V_R(x^{(k)})$. Indeed, for the first term of $u^{(k)}$, we observe that $H^T S^T S H x \geq 0$ for all $x \geq 0$, while the nonnegative term $V_R(x)$ of the decomposition of the vector $\beta A^T y^{(k+1)}$ can be obtained by a non expensive recurrence formula detailed in [2].

4. Numerical experiments

In this section we report some preliminary results obtained by SPDHG method applied to the problem of image super-resolution. Our aim is to evaluate its best performance in terms of reconstruction accuracy and efficiency and compare our results with those obtained by a few representative state-of-the-art algorithms for image super-resolution. To this purpose we select four representative SISR approaches such as Bi-cubic Interpolation, TIP-ASDS-IR [7], ELAD-TIP14 [15], NCSR [8] and perform the image reconstructions under the same testing conditions

used for SPDHG method. The source codes of TIP-ASDS-IR, ELAD-TIP14 and NCSR methods were downloaded from their relative homepages while the Bi-cubic Interpolation method is realized by means of the Matlab function `imresize()`.
We focus here on two test images relative to synthetic and natural images respectively. The Low Resolution (LR) test images are obtained by High Resolution (HR) images (256×256 pixels), normalized in the range [0,1],(see figure 1 (a)) . The HR images are blurred by a 7×7 Gaussian kernel with standard deviation 1.6 and then down-sampled by a scaling factor equal to 3 both in the horizontal and vertical directions. All the numerical computations are performed using MATLAB R2016b on an Intel i7-3770 CPU with 16 GB RAM.
The reconstruction accuracy is measured using the Peak Signal to Noise Ratio parameter (PSNR (dB)) which is reported in table 1 together with the execution times.

We first present the results related to noiseless LR images. The Bi-cubic method is very fast but always obtains worse quality results, therefore it is reported only to appreciate the improvement obtained by the other more complex methods. We observe that, in the case of synthetic images, SPDHG method reaches the best results both in accuracy and execution times (first row of table 1). In the case of natural images (second row of table 1), SPDHG has better quality and execution times compared to TIP-ASDS-IR and ELAD-TIP14 while it has lower PSNR than NCSR. However NCSR results to be the most expensive from the computational point of view. The reconstructed HR images are shown in figure 1.We do not show the images for TIP-ASDS-IR since the PSNR are very close to ELAD-TIP14 while the computation times are larger.

We consider now the case of noisy data, obtained by adding Gaussian white noise with variance $\sigma = 0.005$. We report in table 2 the results obtained with ELAD-TIP14, NCSR and SPDHG, that gave the best results on noiseless data. Figure 2 shows a detail of each reconstructed image. We highlight that the proposed SPDHG method gets the best PSNR values in the shortest time.

We can conclude that SPDHG is a very promising method for fast and accurate super-resolution reconstructions. It allows us to obtain PSNR values better or comparable to those obtained by the best state-of-the-art methods and requires always the lowest computational times.

Image	Bicubic	TIP-ASDS-IR	ELAD-TIP14	NCSR	SPDHG
Synthetic	25.90	30.27 (108.20)	30.48 (8.67)	31.15 (154.29)	**31.8 (3.01)**
Satellite	23.02	25.50 (114.96)	25.42 (8.59)	**26.03** (164.80)	25.7 (**8.18**)

Table 1: PSNR and execution times (in seconds) for each method: best results in bold.

Image	ELAD-TIP14	NCSR	SPDHG
Synthetic	30.00 (8.9)	29.26 (140)	**31.4 (1.8)**
Satellite	25.28 (8.54)	25.60 (149)	**25.61 (4.72)**

Table 2: PSNR and execution times (in seconds) for the best methods on noisy data ($\sigma = 5 \cdot 10^{-3}$): best results in bold.

Acknowledgments
This work has been partially supported by the Italian Institute GNCS - INdAM.

Figure 1: Original and reconstructed HR Images

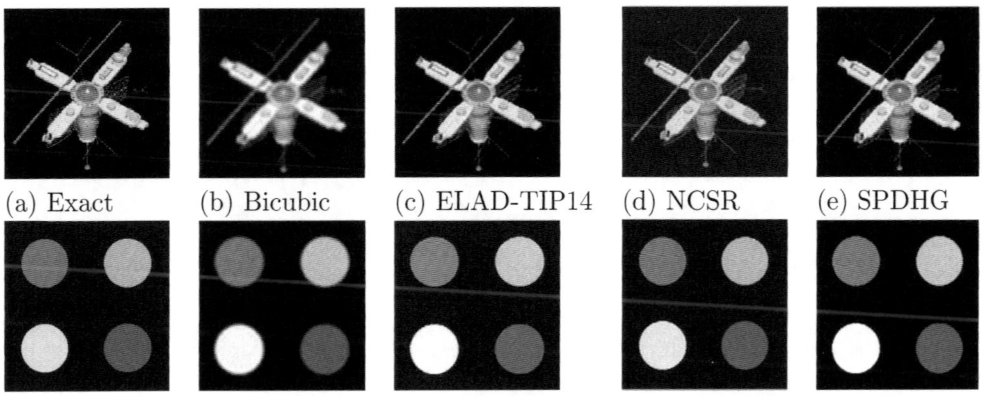

(a) Exact (b) Bicubic (c) ELAD-TIP14 (d) NCSR (e) SPDHG

(a) Exact (b) Bicubic (c) ELAD-TIP14 (d) NCSR (e) SPDHG

Figure 2: Reconstructed HR image details: noisy case.

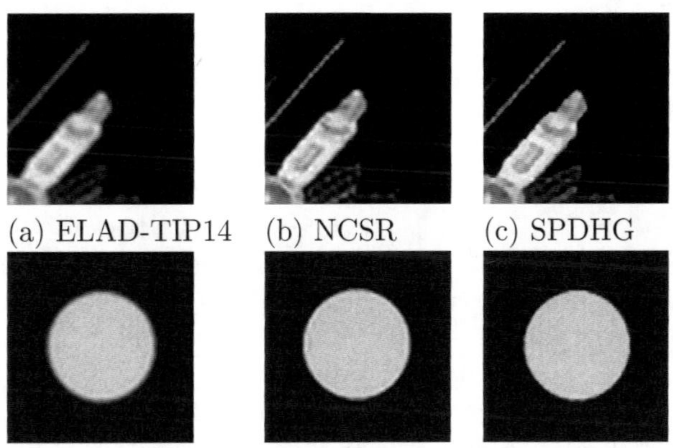

(a) ELAD-TIP14 (b) NCSR (c) SPDHG

(a) ELAD-TIP14 (b) NCSR (c) SPDHG

References

[1] Bertero M., Lanteri H. and Zanni L. 2008 *in Mathematical Methods in Biomedical Imaging and Intensity-Modulated Radiation Therapy (IMRT)* 3763.
[2] Bonettini S., Benfenati A., and Ruggiero V. 2016 *SIAM Journal on Optim.* **26:3** 1741-1772
[3] Bonettini S. and Ruggiero V. 2012 *JMIV* **44**, 236-253
[4] Chambolle A. and Pock T. 2010 *JMIV* **40** 120-145.
[5] Chen H., He X., Teng Q. and Ren C. 2016 *Sig. Proc.: Image Commun.* **43** 68-81.
[6] Coli V. L., Ruggiero V. , Zanni L. 2016 *AIP Conference Proceedings* ,**1776** 040002
[7] Dong W. , Zhang L., Shi G., and Wu X. 2011 *IEEE Transactions on Image Processing*, **20(7)** 1838–1857.
[8] Dong W. , Zhang L., Shi G., and Li X. 2013 *IEEE Transactions on Image Processing* **22(4)** 1620–1630.
[9] Goffin J.L. and Kiwiel K.C. 1999 *Mathematical Programming* **85** 207-211.
[10] Jiang J., Ma X. , Chen C., Lu T., Wang Z., and Ma J. 2017 *IEEE Transaction on Multimedia* **19(1)** 15–26.
[11] Landi G. , Loli Piccolomini E. 20134 *Num. Alg.* **62** 487-504
[12] Marquina A., Osher S. 2008 *J. Sci. Comp.* **37** 367-382.
[13] Montefusco L. and Lazzaro D. 2012 *IEEE Transactions on Image Processing* **21(4)** 1676-1686.
[14] Osher S., Burger M., Goldfarb D., Xu J. and Jin W. 2005 *Mult. Mod. Simul.***4** 460-489.
[15] Peleg T. and Elad M. 2014 *IEEE Transactions on Image Processing* **23(6)** 2569–2582.
[16] Zanella R., Boccacci P., Zanni L., Bertero M. 2009 *Inverse Probl.* **25** 045010.
[17] Zhu M. and Chan T., 2008 *UCLA CAM Report* [08-34]

IOP Conf. Series: Journal of Physics: Conf. Series **904** (2017) 012010 doi:10.1088/1742-6596/904/1/012010

Minimum Mean Square Distance Estimation of Subspaces in presence of Gaussian sources with application to STAP detection

R. Ben abdallah [1], A. Breloy [2], M. N. El Korso [3], D. Lautru [4], H. Hafdallah Ouslimani [5]

LEME, Universite Paris Ouest, 50, rue de Sevres, 92410 Ville dAvray, France

E-mail: [1]rayen.benabdallah@u-paris10.fr, [2]abreloy@u-paris10.fr, [3]m.elkorso@u-paris10.fr, [4]dlautru@u-paris10.fr, [5]houslima@u-paris10.fr

Abstract. We consider the problem of signal subspace estimation from a given sample set. We rely on the Bayesian framework developped in [1] to obtain minimum mean square distance (MMSD) estimators, which minimize the expected distance between the true projection matrix $\mathbf{U}\mathbf{U}^T$ and its estimate $\widehat{\mathbf{U}}\widehat{\mathbf{U}}^T$. In this work, we extend the estimators of [1] to the context of linear model with Gaussian sources, with respectively Bingham and von Mises Fisher priors for the basis $\bar{\mathbf{U}}$. Numerical simulations are given in order to assess the performance of the proposed estimator. The interest of the considered approach is finally illustrated on a real data set, for a Space Time Adaptive Processing (STAP) application.

Index Term Subspace estimation, minimum mean square distance estimation, Bayesian setting, Bingham distribution, von Mises Fisher distribution, STAP detection.

1. Introduction

Subspace estimation is an ubiquitous issue in signal processing. It aims at estimating the low dimensional subspace where information (usually signals of interest) lies in. This problem arises in numerous applications, such as in PCA, DOA estimation [2], interference cancellation [11] or signal detection [10].

In the literature, a widely used model is the linear model where the data matrix is modeled as $\mathbf{Y} = \mathbf{US} + \mathbf{N}$, in which $\mathbf{U}$ spans the subspace of interest, $\mathbf{S}$ is the signal of interest and $\mathbf{N}$ is an additive noise. The addressed problem is to estimate $\mathbf{U}$ from the observations $\mathbf{Y}$. The most usual mean of estimating this subspace $\mathbf{U}$ is the singular value decomposition (SVD) of the sample covariance matrix (SCM), i.e., taking the largest eigenvectors of the SCM. This provides usually an accurate estimator of range space $\mathcal{R}(\mathbf{U})$ of $\mathbf{U}$ for high signal to noise ratio (SNR) and/or for large number of sample support contexts. However, this method shows its limits outside these asymptotic regimes. A possible solution to overcome this issue is to incorporate some prior knowledge into the estimation process, as proposed in [1], [5]. For example, in a Bayesian setting, the minimum mean square error (MMSE) [5] which minimizes the average Euclidean distance between the true subspace $\mathbf{U}$ and its estimate $\widehat{\mathbf{U}}$, and the minimum mean square distance (MMSD) [1] which aims to minimize the distance between the true projection matrix $\mathbf{U}\mathbf{U}^T$ and its estimate $\widehat{\mathbf{U}}\widehat{\mathbf{U}}^T$, were proposed. More specially, in [1] the MMSD is derived for both linear and covariance models with uniform prior distributions for $\mathbf{S}$. In this paper, we extend the work of [1] for the linear model, for the case where, the signal model $\mathbf{S}$ is assumed to follow a Gaussian law and for $\mathbf{U}$ which has respectively a Bingham and a von Mises Fisher prior distributions as [1].

In this work, we extend the estimators of [1] to the context of linear model with Gaussian sources, with respectively Bingham and von Mises Fisher priors for the basis $\mathbf{U}$. Numerical simulations are given in order to assess the performance of the proposed estimator. The interest of the considered approach is finally illustrated on a real data set, for a Space Time Adaptive Processing (STAP) application.

2. Theory background and model setup

Along this paper, we denote by $\mathbf{Y} \in \mathbb{R}^{N\times K}$ the data matrix, $\mathbf{U} \in \mathbb{R}^{N\times P}$ the unknown subspace of interest, $\mathbf{S} \in \mathbb{R}^{P\times K}$ the signal of interest and $\mathbf{N} \in \mathbb{R}^{N\times K}$ the additive noise, in which, N

IOP Conf. Series: Journal of Physics: Conf. Series **904** (2017) 012010 doi:10.1088/1742-6596/904/1/012010

denotes the number of antennas, K the number of snapshots, P the dimension of subspace of interest ($P \ll N$) and $p(\mathbf{Y} \mid \mathbf{U})$ the conditional probability of $\mathbf{Y}$ given $\mathbf{U}$. Tr{.} stands for trace of a given matrix and E {.} is the expectation of a given random variable.

2.1. Minimum mean square distance (MMSD)

The MMSD estimator minimizes the distance between the range spaces $\mathcal{R}(\widehat{\mathbf{U}})$ and $\mathcal{R}(\mathbf{U})$, i.e., $\| \widehat{\mathbf{U}}\widehat{\mathbf{U}}^T - \mathbf{U}\mathbf{U}^T \|_F^2$ which is a natural metric in Grassman manifold $G_{N,P}$ (the set of P-dimensional subspace in $\mathbb{R}^N$) as proven in [3], [4]. This estimator is therefore expressed as: $\widehat{\mathbf{U}}_{\text{mmsd}} = \arg\min_{\widehat{\mathbf{U}}} \| \widehat{\mathbf{U}}\widehat{\mathbf{U}}^T - \mathbf{U}\mathbf{U}^T \|_F^2 = \arg\max_{\widehat{\mathbf{U}}} \mathrm{E}\{\mathrm{Tr}\{\widehat{\mathbf{U}}^T\mathbf{U}\mathbf{U}^T\widehat{\mathbf{U}}\}\}$.As shown in [1], the MMSD estimator is obtained in general from the P largest eigenvectors of the matrix $\mathbf{M}(p(\mathbf{U} \mid \mathbf{Y})) = \int \mathbf{U}\mathbf{U}^T p(\mathbf{U} \mid \mathbf{Y})d\mathbf{U}$ [1] and [5] which is expressed below as:

$$\widehat{\mathbf{U}}_{\text{mmsd}} = \mathcal{P}_P\left\{\int \mathbf{U}\mathbf{U}^T p(\mathbf{U} \mid \mathbf{Y})d\mathbf{U}\right\} = \mathcal{P}_P\left\{\mathbf{M}(p(\mathbf{U} \mid \mathbf{Y}))\right\} \tag{1}$$

where, $\mathcal{P}_P\{.\}$ is an operator that stands for the P strongest eigenvectors of a given matrix. The generic expression of the MMSD depends on $p(\mathbf{U} \mid \mathbf{Y})$ which has to be specified based on both the data model and the prior distribution of $\mathbf{S}$ and $\mathbf{U}$.

2.2. Model setup

In this paper, we assume that the data is the sum of a signal of interest and a zero white Gaussian noise. Furthermore, we assume that the columns of the additive noise $\mathbf{N}$ are i.i.d. and each follows $\mathcal{N}(0, \sigma^2\mathbf{I})$ distribution and the columns of the signal of interest $\mathbf{S}$ are i.i.d. and each follows $\mathcal{N}(0, \sigma_S^2\mathbf{I})$ distribution. By contrast, we note that in [1], $\mathbf{S}$ is assumed to follow a uniform law. The linear Gaussian model is given by:

$$\mathbf{Y} = \mathbf{U}\mathbf{S} + \mathbf{N} \tag{2}$$

Since, we adopt a Bayesian framework in this paper, a prior distribution must be assigned to $\mathbf{U}$. With regarding to Grassman and Stiefel manifold, the most known distributions are Bingham and von Mises Fisher (vMF) [7], [8] which are specified below.

2.2.1. General Bingham von Mises Fisher distribution (BMF)

The BMF distribution has as parameter matrices $\mathbf{A} \in \mathbb{R}^{N\times N}$, $\mathbf{B} \in \mathbb{R}^{P\times P}$ and $\mathbf{C} \in \mathbb{R}^{N\times P}$ [6]. In this paper, it is sufficient to assume that $\mathbf{B}$ is a diagonal matrix, $\mathbf{A}$ is a symmetric matrix, $\mathbf{C}$ is an orthogonal matrix. The matrix $\mathbf{U}$ is said to have $\mathbf{U} \sim BMF(\mathbf{A}, \mathbf{B}, \mathbf{C})$ distribution, if its probability density function is given as below : $p_{\mathrm{BMF}}(\mathbf{U}) \propto \mathrm{etr}\{\mathbf{C}^T\mathbf{U} + \mathbf{B}\mathbf{U}^T\mathbf{A}\mathbf{U}\}$ in which, etr{.} stands for the exponential of the trace of a given matrix and $\propto$ indicates "proportional to". The Bingham and vMF are special cases of this distribution.

2.2.2. Bingham distribution

The Bingham distribution [7] is a particular case of the BMF distribution where, $\mathbf{C}$ is a null matrix. The matrix $\mathbf{U}$ is said to have a Bingham distribution with parameters $\mathbf{A}$ and $\mathbf{B}$, i.e., $\mathbf{U} \sim B(\mathbf{A}, \mathbf{B})$, if its probability density function reads as $p_{\mathrm{B}}(\mathbf{U}) \propto \mathrm{etr}\{\mathbf{B}\mathbf{U}^T\mathbf{A}\mathbf{U}\}$in which, $\mathbf{A}$ gathers the prior knowledge about the subspace of interest. Here, we focus on the case where $\mathbf{A}$ is proportional to a projection matrix spanned by the basis $\bar{\mathbf{U}}$. Thus, we denote $\mathbf{B} = \mathbf{I}$ and $\mathbf{A} = \kappa\bar{\mathbf{U}}\bar{\mathbf{U}}^T$:

$$p_{\mathrm{B}}(\mathbf{U}) \propto \mathrm{etr}\{\kappa\mathbf{U}^T\bar{\mathbf{U}}\bar{\mathbf{U}}^T\mathbf{U}\} \tag{3}$$

where $\bar{\mathbf{U}}$ is therefore the center of the distribution and κ is the concentration parameter of the distribution.

2.2.3. vMF distribution

The vMF distribution [8] also is a special case of BMF for which $\mathbf{A}$ and $\mathbf{B}$ are nulls. $\mathbf{U}$ is said to have a vMF distribution with parameters $\mathbf{C}$, i.e., $\mathbf{U} \sim vMF(\mathbf{C})$, if its probability density function is given by: $p_{\mathrm{vMF}}(\mathbf{U}) \propto \mathrm{etr}\{\mathbf{C}^T\mathbf{U}\}$. As previously, we focus on vMF distribution where $\mathbf{C}$ is proportional to a unitary matrix $\bar{\mathbf{U}}$. We denote $\mathbf{C} = \kappa\bar{\mathbf{U}}$, thus,

$$p_{\mathrm{vMF}}(\mathbf{U}) \propto \mathrm{etr}\{\kappa\mathbf{U}^T\bar{\mathbf{U}}\} \tag{4}$$

Remark1: Note that the Bingham distribution p_{B} characterizes the distribution of $\mathcal{R}(\mathbf{U})$ range space of $\mathbf{U}$. The vMF distribution p_{vMF} characterizes the distribution of the orthogonal base $\mathbf{U}$. Hence, as defined in (3) and (4), p_B is invariant by rotation while, p_{vMF} is not.

3. Proposed MMSD estimators in presence of Gaussian sources

Here, we develop the MMSD estimator from 2.1 for the model specified in 2.2 using the prior distributions of $\mathbf{U}$ given in 2.2. Let us derive the conditional probability of $\mathbf{Y}$ given $\mathbf{U}$:

$$\begin{aligned} p(\mathbf{Y} \mid \mathbf{U}) &= \int p(\mathbf{Y} \mid \mathbf{S}, \mathbf{U})p(\mathbf{S})d\mathbf{S} \propto \int \mathrm{etr}\left\{-\frac{1}{2\sigma^2}(\mathbf{Y}-\mathbf{US})^T(\mathbf{Y}-\mathbf{US})\right\}\mathrm{etr}\{-\frac{1}{2\sigma_{\mathbf{S}}{}^2}\mathbf{S^TS}\}d\mathbf{S} \\ &\propto \mathrm{etr}\left\{\alpha_G\mathbf{Y}^T\mathbf{U}\mathbf{U}^T\mathbf{Y} - \frac{1}{2\sigma^2}\mathbf{Y}^T\mathbf{Y}\right\} \end{aligned} \tag{5}$$

where $\alpha_G = \frac{\sigma_{\mathbf{S}}{}^2}{2\sigma^2(\sigma^2+\sigma_{\mathbf{S}}{}^2)}$. In the next subsection, we develop the MMSD estimators, for Bingham and vMF prior distributions on $\mathbf{U}$.

3.1. Bingham prior

We start by deriving the posterior probability of $\mathbf{U}$, which is given by: $p_{\text{B-G}}(\mathbf{U} \mid \mathbf{Y}) \propto p(\mathbf{Y} \mid \mathbf{U})p_{\mathrm{B}}(\mathbf{U}) \propto \mathrm{etr}\left\{\mathbf{U}^T\left[\alpha_G\mathbf{Y}\mathbf{Y}^T + \kappa\bar{\mathbf{U}}\bar{\mathbf{U}}^T\right]\mathbf{U}\right\}$, consequently, the posterior probability $p_{\text{B-G}}(\mathbf{U} \mid \mathbf{Y}) \propto B(\alpha_G\mathbf{Y}\mathbf{Y}^T + \kappa\bar{\mathbf{U}}\bar{\mathbf{U}}^T, \mathbf{I})$. Then, we move to the derivation of the MMSD estimator using (1), we obtain: $\widehat{\mathbf{U}}_{\text{mmsd-B-G}} = \mathcal{P}_P\left\{\int \mathbf{U}\mathbf{U}^T p_{\text{B-G}}(\mathbf{U} \mid \mathbf{Y})d\mathbf{U}\right\} = \mathcal{P}_P\left\{\int \mathbf{U}\mathbf{U}^T \mathrm{etr}\left\{\mathbf{U}^T\left[\alpha_G\mathbf{Y}\mathbf{Y}^T + \kappa\bar{\mathbf{U}}\bar{\mathbf{U}}^T\right]\mathbf{U}\right\}d\mathbf{U}\right\}$. At this point, we use proposition 1 of [1] which states that $\int \mathbf{U}\mathbf{U}^T\mathrm{etr}\{\mathbf{U}^T(\alpha_G\mathbf{Y}\mathbf{Y}^T + \kappa\bar{\mathbf{U}}\bar{\mathbf{U}}^T)\mathbf{U}\}d\mathbf{U}$ and $\alpha_G\mathbf{Y}\mathbf{Y}^T + \kappa\bar{\mathbf{U}}\bar{\mathbf{U}}^T$ share same eigenvectors and its correspondents eigenvalues have the same order. Consequently, we obtain:

$$\widehat{\mathbf{U}}_{\text{mmsd-B-G}} = \mathcal{P}_P\left\{\alpha_G\mathbf{Y}\mathbf{Y}^T + \kappa\bar{\mathbf{U}}\bar{\mathbf{U}}^T\right\} \tag{6}$$

3.2. vMF prior

We start with deriving the posterior probability of $\mathbf{U}$ regarding the vMF prior: $p_{\text{vMF-G}}(\mathbf{U} \mid \mathbf{Y}) \propto p(\mathbf{Y} \mid \mathbf{U})p_{\mathrm{vMF}}(\mathbf{U}) \propto \mathrm{etr}\left\{\kappa\bar{\mathbf{U}}^T\mathbf{U} + \alpha_G\mathbf{U}^T\mathbf{Y}\mathbf{Y}^T\mathbf{U}\right\}$. In this case, $p_{\text{vMF-G}} \propto BMF(\mathbf{Y}\mathbf{Y}^T, \alpha_G\mathbf{I}, \kappa\bar{\mathbf{U}})$. Then, we use (1) to derive the estimator: $\widehat{\mathbf{U}}_{\text{mmsd-vMF-G}} = \mathcal{P}_P\left\{\int \mathbf{U}\mathbf{U}^T p_{\text{vMF-G}}(\mathbf{U} \mid \mathbf{Y})d\mathbf{U}\right\} = \mathcal{P}_P\left\{\int \mathbf{U}\mathbf{U}^T\mathrm{etr}\left\{\kappa\bar{\mathbf{U}}^T\mathbf{U} + \alpha_G\mathbf{U}^T\mathbf{Y}\mathbf{Y}^T\mathbf{U}\right\}d\mathbf{U}\right\}$. Unfortunately, there is no closed form of the the MMSD in this context. However, an efficient Gibbs simpling can be generated to draw the BMF distribution [6]. In order to compute the so-called induced arithmetic mean (IAM) [9] of the unitary matrix $\mathbf{U}^{(n)} \sim BMF(\mathbf{Y}\mathbf{Y}^T, \alpha_G\mathbf{I}, \kappa\bar{\mathbf{U}})$ and then, the MMSD is given by:

$$\widehat{\mathbf{U}}_{\text{mmsd-vMF-G}} = \mathcal{P}_P\left\{\frac{1}{N_r}\sum_{n=N_{bi}+1}^{N_{bi}+N_r}\mathbf{U}^{(n)}\mathbf{U}^{(n)T}\right\} \tag{7}$$

where N_{bi} stands for the burn-in samples which is the number of thrown samples from the Markov chain, N_r is the number of samples used to approximate the estimator.

Discussion: We recall the expressions of the derived estimators in linear model context in [1], where $\mathbf{S}$ has a prior uniform distribution. For Bingham prior, the estimator is given by: $\widehat{\mathbf{U}}_{\text{mmsd-B-U}} = \mathcal{P}_P\left\{\alpha_U\mathbf{Y}\mathbf{Y}^T + \kappa\bar{\mathbf{U}}\bar{\mathbf{U}}^T\right\}$. For the vMF prior, the MMSD estimator is obtained as (7) where $\mathbf{U}^{(n)} \sim BMF(\mathbf{Y}\mathbf{Y}^T, \alpha_U\mathbf{I}, \kappa\bar{\mathbf{U}})$ and $\alpha_U = \frac{1}{2\sigma^2}$.

The main difference between our MMSD estimators (6), (7) and those presented previously [1] lies in the scalar α_G from (6). Notice that for low SNR, $\alpha_G \ll \alpha_U$ which means that our estimators are more shifted towards the prior $\bar{\mathbf{U}}$ than the ones given above. While, for high SNR, $\alpha_G \approx \alpha_U$ which means that the estimators coincide, as shown in fig. 1.

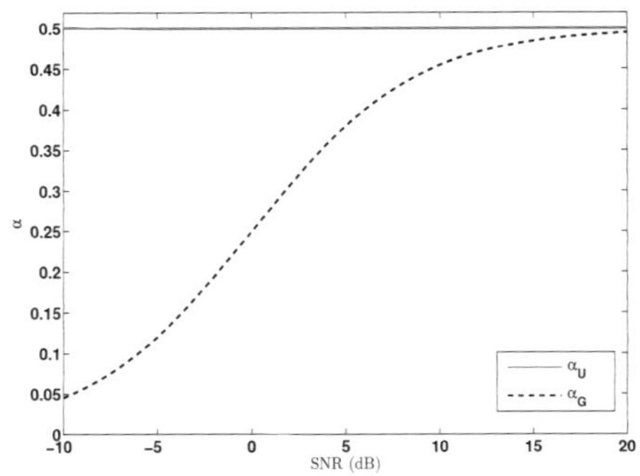

Figure 1. $\alpha_G = \frac{\sigma_S{}^2}{2\sigma^2(\sigma^2+\sigma_S{}^2)}$ and $\alpha_U = \frac{1}{2\sigma^2}$ in function of SNR for $\sigma^2 = 1$.

4. Simulation

In this section, we illustrate the performance of the proposed estimator through evaluating the average fraction of energy (AFE). The AFE of a given estimator $\widehat{\mathbf{U}}$ is expressed as: $\text{AFE}(\widehat{\mathbf{U}}) = \text{E}\{\text{Tr}\{\mathbf{U}^T\widehat{\mathbf{U}}\widehat{\mathbf{U}}^T\mathbf{U}\}\}$ which is considered as a criteria of performance, since, it evaluates the closeness of an estimator towards the random generated matrix $\mathbf{U}$. We set K=5, N=20, P=5, SNR=0dB, κ=20 and $\bar{\mathbf{U}} = \begin{bmatrix}\mathbf{I}_p \\ 0\end{bmatrix}$. We compare in this part the proposed MMSD estimators denoted $\widehat{\mathbf{U}}_{\text{mmsd-B-G}}$ and $\widehat{\mathbf{U}}_{\text{mmsd-vMF-G}}$ with those presented in [1] denoted respectively $\widehat{\mathbf{U}}_{\text{mmsd-B-U}}$ and $\widehat{\mathbf{U}}_{\text{mmsd-vMF-U}}$ and with the SVD estimator of the SCM, denoted $\widehat{\mathbf{U}}_{\text{SCM}}$ which takes no prior knowledge of subspace, i.e., $\mathbf{U}$ is deterministic. For computing the IAM, we fix $N_{bi} = 10$ and $N_r = 200$. Then, we illustrate figures below which examine the impact of SNR and K on the performance of estimators and for Gaussian prior for $\mathbf{S}$.

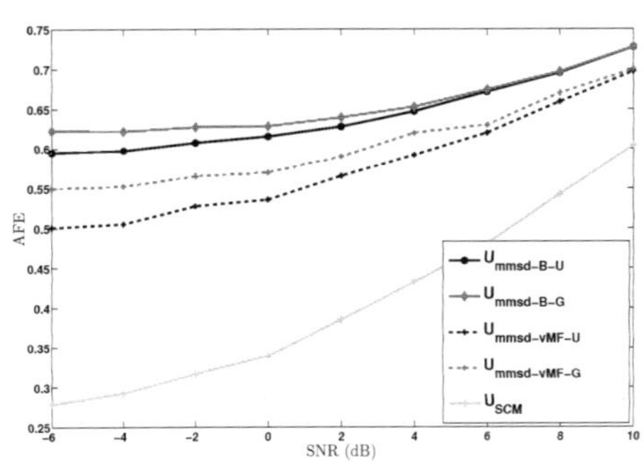

Figure 2. Fraction of energy of $\widehat{\mathbf{U}}$ in $\mathcal{R}(\mathbf{U})$ versus SNR, $K = 5$, Gaussian prior for $\mathbf{S}$.

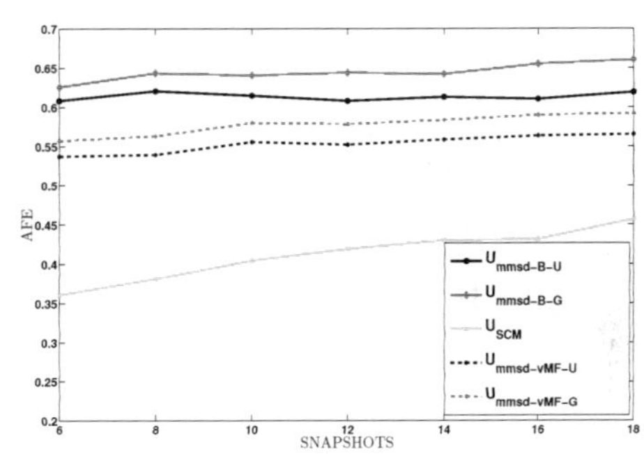

Figure 3. Fraction of energy of $\widehat{\mathbf{U}}$ in $\mathcal{R}(\mathbf{U})$ versus K, for $SNR = 0dB$, Gaussian prior for $\mathbf{S}$.

According to fig.2 and fig.3, the SVD estimator shows his poor performance outside the asymptotic regimes. In regard to fig.2 and fig.3, we notice that for low SNR and/or small number of snapshots, the proposed estimators (6), (7), prove a better performance compared respectively to those presented in discussion above. In asymptotic regimes, the proposed estimators have almost same performance as the ones given in discussion above.

5. Application to STAP low rank detection

In this section, we illustrate the interest of MMSD estimators for space-time adaptive processing (STAP) on a real data set, provided by the French agency DGA/MI [16]. STAP [12] is applied to airborne radar in order to detect moving targets. Typically, the radar receiver consists in an array of Q antenna elements processing M pulses in a coherent processing interval ($N_c = MQ$). In the classical framework [15], we assume that the $K+1$ samples are drawn as:

$$\begin{cases} \mathbf{z}_0 &= \mathbf{d} + \mathbf{c}_0 + \mathbf{n}_0 \\ \mathbf{z}_k &= \mathbf{c}_k + \mathbf{n}_k, \forall k \in \{1...K\} \end{cases} \tag{8}$$

where the tested cell $\mathbf{z}_0$ potentially contains a target $\mathbf{d} = \alpha_0 \mathbf{p}(\theta, v_t)$ to be detected. α_0 is the amplitude and $\mathbf{p}$ is the normalized steering vector, v_t is the target speed and θ is the DOA of the target (α_0, v_t and θ are unknown). For each sample, the additive noise is the sum of the ground clutter $\mathbf{c}$ and white Gaussian noise $\mathbf{n}$. The secondary data $\{\mathbf{z}_k\}$ are assumed to be i.i.d., signal and outlier free. Thanks to the Brennan rule [13], the ground clutter is known to be contained in a low dimensional subspace $\mathbf{\Pi}_c = \mathbf{U}_c \mathbf{U}_c^H$ of (known) rank R. Hence, if we concatenate the secondary data, we obtain in matrix form:

$$\mathbf{Z} = \mathbf{\Pi}_c \mathbf{C} + \mathbf{N} \tag{9}$$

It is worth mentioning that, even if the fluctuation of the clutter $\mathbf{c}$ may not be white Gaussian in practice, results reveal that the considered framework remains robust to such assumptions (c.f. Figure 4). We investigate the performance of the low rank adaptive normalized matched filter (ANMF) detector [12]:

$$\widehat{\mathbf{\Lambda}}(\widehat{\mathbf{\Pi}}_c) = \frac{|\ \mathbf{d}^H \widehat{\mathbf{\Pi}}_c^\perp \mathbf{z}_0\ |^2}{|\ \mathbf{d}^H \widehat{\mathbf{\Pi}}_c^\perp \mathbf{d}\ ||\ \mathbf{z}_0^H \widehat{\mathbf{\Pi}}_c^\perp \mathbf{z}_0\ |} \tag{10}$$

where $\widehat{\mathbf{\Pi}}_c = \widehat{\mathbf{U}}_c \widehat{\mathbf{U}}_c^T$ is a clutter subspace projector (CSP) estimated from the secondary data $\{\mathbf{z}_k\}$. The performance of a given detector depends on the accuracy of the CSP estimator from which it is built. This is why we compare the performance of the following detectors: $\widehat{\mathbf{\Lambda}}_{\text{SCM}}$, where the CSP is estimated from the SVD of the SCM. $\widehat{\mathbf{\Lambda}}_{\text{SFPE}}$, where the CSP is estimated from the SVD of the shrinkage fixed point estimator with minimal regularization parameter γ [14]. $\widehat{\mathbf{\Lambda}}_{\text{mmsd-B-G}}$, where the CSP is estimated using our proposed schemes thanks to (6). Since the parameters α_G and κ are here unknown, we will illustrate the performance of the estimator defined as $\widehat{\mathbf{U}}_{\text{mmsd-B-G}} = \mathcal{P}_R\left[\mathbf{Z}\mathbf{Z}^H + \beta\bar{\mathbf{U}}\bar{\mathbf{U}}^H\right]$ for different values of the ratio parameter $\beta = \kappa/\alpha_G$. The prior $\bar{\mathbf{U}}$ is computed from the SVD of the STAP covariance matrix model in [12]. $\mathbf{\Lambda}_{\text{prior}}$, the (non adaptive) detector using the prior $\bar{\mathbf{U}}$ from the SVD of the STAP covariance matrix model in [12].

Figure 4 displays the output of the different detectors (i.e. the value of the ANMF for a grid of target parameters) for a configuration where the tested cell $\mathbf{z}_0$ contains 10 targets and the number of secondary data is limited but still larger than the evaluated clutter rank ($2R < K = 100 < N_c/2$). We observe that, in this context, the detector $\widehat{\mathbf{\Lambda}}_{\text{SFPE}}$ allows for target detection with a relatively small false alarm rate. The detector $\widehat{\mathbf{\Lambda}}_{\text{mmsd-B-G}}$ represents a trade off between $\widehat{\mathbf{\Lambda}}_{\text{SCM}}$ (i.e. $\beta = 0$) and $\widehat{\mathbf{\Lambda}}_{\text{prior}}$ ($\beta \to \infty$). As observed, $\widehat{\mathbf{\Lambda}}_{\text{SCM}}$ does not allows for target detection since the target responses seems also canceled. This is probably due to the spreading of targets in the secondary data and the poor robustness of the SCM. The fully deterministic detector $\widehat{\mathbf{\Lambda}}_{\text{prior}}$ does not cancel the ground clutter as we can only observe false alarms on the ridge and no distinguishable targets. However, the detector $\widehat{\mathbf{\Lambda}}_{\text{mmsd-B-G}}$ provides a trade off that allows for interference rejection and reliable target detection. This detector even outperforms $\widehat{\mathbf{\Lambda}}_{\text{SFPE}}$ for appropriate values of the parameter β, which illustrates the interest of introducing some prior information in an adaptive subspace estimation process.

As prospects of this work, it will be interesting to develop adaptive methods for the selection of the parameter $\beta = \kappa/\alpha_G$. A second point to be investigated is the development of MMSD estimators that account for others signal models (e.g. covariance models, colored and non-Gaussian sources).

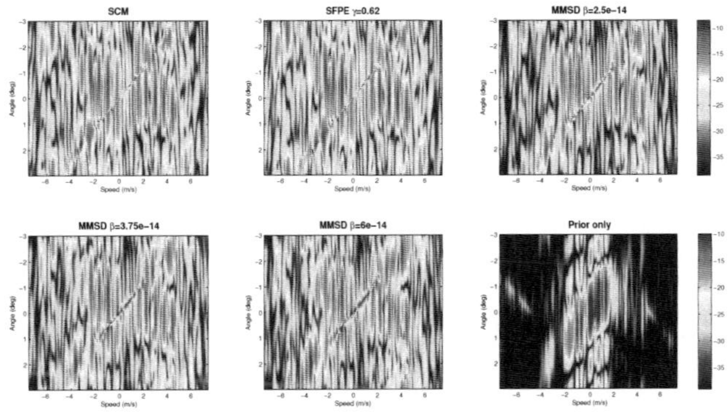

Figure 4. Output of the ANMF for a grid of target parameters (θ, v_t) where $\theta \in$ [-3, 3] and $v_t \in$[-6, 6] for a given detector. The represented detectors are $\widehat{\mathbf{\Lambda}}_{SCM}$, $\widehat{\mathbf{\Lambda}}_{SFPE}$, $\widehat{\mathbf{\Lambda}}_{\text{mmsd-B-G}}$ for several values of β, $\mathbf{\Lambda}_{\text{prior}}$. the number of secondary data is K=100. The number of sensors is $Q = 4$ and the number of coherent pulses is $M = 64$, the size of the data is then $N_c = QM = 256$. The center frequency and the bandwidth are respectively equal to f_0 = 10GHz and the bandwidth $B = 5MHz$. The radar celerity is V = 100m/s. The inter-element spacing is $d = 0,3$m and the pulse repetition frequency is f_r = 1kHz. The clutter rank, computed from Brennan Rule [13], is $R = 45$ and the CNR is evaluated around 20dB.

References

[1] O. Besoon, N. Dobigeon and J. Tourneret, Minimum mean square distance estimation of a subspace, IEEE Trans., Signal Process., vol. **59**, No. 12, pp 5709-5720, December 2011.

[2] R. O. Schimidt, Multiple emitter location and signal parameter estimation, IEEE Trans., Antennas Propag., vol. **34**, no 3 , pp 276-280, March 1986.

[3] A. Edelman, T. Arias, and S. Smith, The geometry of algorithms with orthogonality constraints, SIAM J. Matrix Anal. Appl. , vol. **20**, no. 2, pp. 303-353, 1998.

[4] G. Golub and C. V. Loan, Matrix Computations, 3rd ed. Baltimore, MD: Johns Hopkins Univ. Press, 1996.

[5] A. Srivastava, A Bayesian approach to geometric subspace estimation, IEEE Trans. Sig. Process., vol. **48**, no. 5, pp. 1390-1400, May 2000.

[6] P. D. Hoff, Simulation of the matrix Bingham-von Mises-Fisher distribution with applications to multivariate and relational data, J. Computat. Graph. Statist., vol. **18**, no. 2, pp. 438-456, June 2009.

[7] K. V. Mardia and P. E. Jupp, Directional Statistics. New York: Wiley, pp. 167-183, 1999.

[8] Y. Chikuse, Statistics on Special Manifolds. New York: Springer-Verlag, pp. 27-52, 2003.

[9] A. Sarlette and R. Sepulchre, Consensus optimization on manifolds, SIAM J. Contr. Optimiz., vol. **48**, no. 1, pp. 56-76, 2009.

[10] M. Haardt, M. Pesavento, F. Roemer and M. N. El Korso, Subspace Methods and Exploitation of Special Array Structures, Electro. Refer. in Sig. Process. : Array and Statis. Sig. Process. (M. Viberg, ed.), vol. **3**, pp. 651-717, 2014.

[11] J. Liu, M. S. Shbat and V. Tuzlukov, Interference Cancellation and DOA Estimation by Generalized Receiver Applying LMS and MUSIC Algorithms, Progress In Electromagnetics Research Symposium Proceed., pp 187-190, March 2013.

[12] J. Ward, Space-time adaptive processing for airborne radar, Tech. Rep., Lincoln Lab., MIT, Lexington, Mass., USA, December 1994.

[13] L. E. Brennan and F. M. Staudaher, Subclutter visibility demonstration, Tech. Rep., RL-TR-92-21, Adapt. Sensors Incorporated, March 1992.

[14] F. Pascal, Y. Chitour and Y. Quek, Generalized robust shrinkage estimator and its application to STAP detection problem, Sig. Process., arXiv:1311.6567 , September 2014.

[15] G. Ginolhac, P. Forster, J. P. Ovarlez and F. Pascal, Détecteurs STAP Rang Réduit, Traitement du Signal, vol. **28**, no. 1-2, pp 143-170, April 2011.

[16] J. P. Ovarlez, F. Le Chevalier and S. Bidon, les données de club STAP, Introduction au STAP, Traitement de signal, vol. **28**, 2011

IOP Conf. Series: Journal of Physics: Conf. Series **904** (2017) 012011 doi:10.1088/1742-6596/904/1/012011

Magnetization moment recovery using Kelvin transformation and Fourier analysis

L. Baratchart[1], **J. Leblond**[1], **E. A. Lima**[2], **D. Ponomarev**[3,⋆]

[1] Projet APICS, INRIA, 2004 Route des Lucioles, 06902 Sophia Antipolis Cedex, France
[2] Earth, Atmospheric, and Planetary Sciences department, MIT, Cambridge, MA 02139, USA
[3] Laboratoire POEMS, ENSTA ParisTech, 828 Boulevard des Maréchaux, 91762 Palaiseau Cedex, France
[⋆] Corresponding author

E-mail: dmitry.ponomarev@ensta-paristech.fr

Abstract. In the present work, we consider a magnetization moment recovery problem, that is finding integral of the vector function (over its compact support) whose divergence constitutes a source term in the Poisson equation. We outline derivation of explicit asymptotic formulas for estimation of the net magnetization moment vector of the sample in terms of partial data for the vertical component of the magnetic field measured in the plane above it. For this purpose, two methods have been developed: the first one is based on approximate projections onto spherical harmonics in Kelvin domain while the second stems from analysis in Fourier domain following asymptotic continuation of the data. Recovery results obtained by both methods agree and are illustrated numerically by plotting formulas for net moment components with respect to the size of the measurement area.

1. Introduction and problem formulation

Earth rocks and meteorites may preserve invaluable records of ancient planetary and solar nebula magnetic fields in the form of remanent magnetization. Recent advances in magnetometry (e.g., SQUID microscopy technique) have made it possible to measure magnetic fields of very low intensity generated by some rocks, and extraction of this relict magnetic information has become reality in cases that were previously inaccessible using standard rock magnetometers. An endeavor to develop a robust and efficient method for processing these data leads to a number of challenging problems such as effective extension of the restricted measurement data and extraction of certain features of the magnetization (typically, its mean value) that may still allow for retrieving those primordial records without having to solve the entire inverse problem for the underlying spatial distribution of magnetic sources. In particular, we are concerned with the following setting previously discussed in [2, 3].

Suppose there is a localized sample whose magnetization distribution is stationary and described by an unknown vector function

$$\vec{M}(\vec{x}) \equiv (M_1(\mathbf{x}, x_3), M_2(\mathbf{x}, x_3), M_3(\mathbf{x}, x_3))^T, \quad \mathbf{x} \equiv (x_1, x_2)^T, \quad \vec{x} \equiv (\mathbf{x}, x_3)^T,$$

supported on a subset $Q \subset \mathbb{R}^3$.

IOP Conf. Series: Journal of Physics: Conf. Series **904** (2017) 012011 doi:10.1088/1742-6596/904/1/012011

Representation of the produced magnetic field in terms of the scalar potential Φ leads to the Poisson equation

$$\vec{B}/\mu_0 = -\nabla\Phi + \vec{M} \quad \Rightarrow \quad \Delta\Phi = \nabla\cdot\vec{M},$$

where $\mu_0 = 4\pi\cdot 10^{-7}$ H$\cdot$ m^{-1} is the magnetic constant. Therefore,

$$\Phi(\mathbf{x}, x_3) = -\frac{1}{4\pi}\iiint_{\mathbb{R}^3}\frac{\nabla\cdot\vec{M}(\vec{t})}{\left(|\mathbf{x}-\mathbf{t}|^2+(x_3-t_3)^2\right)^{1/2}}d^3t.$$

In a typical experimental set-up, the vertical component of the magnetic field $B_3 = -\mu_0\frac{\partial}{\partial x_3}\Phi$ is measured on a part of the horizontal plane at height $x_3 = h > 0$

$$\begin{aligned} B_3(\mathbf{x}, h) &= \frac{\mu_0}{4\pi}\iiint_Q \left[3(h-t_3)\left(M_1(\vec{t})(x_1-t_1)+M_2(\vec{t})(x_2-t_2)\right)\right. \\ &\left. + M_3(\vec{t})\left(2(h-t_3)^2-|\mathbf{x}-\mathbf{t}|^2\right)\right]\frac{d^3t}{\left(|\mathbf{x}-\mathbf{t}|^2+(h-t_3)^2\right)^{5/2}}. \end{aligned} \tag{1}$$

Then, knowing $B_3(\mathbf{x}, h)$ on some subset of the measurement plane above the sample, the main physical interest is to estimate the net magnetization moment vector of the sample

$$\vec{m} \equiv (m_1, m_2, m_3)^T := \iiint_Q \vec{M}(\vec{x})\, d^3x \in \mathbb{R}^3.$$

In the present work, we assume that this subset is a disk $D_A := \left\{\mathbf{x}\in\mathbb{R}^2 : |\mathbf{x}| \le A\right\}$.

2. Kelvin transform approach

2.1. Motivation of the method

Let us imagine the situation when we have available data for the potential on a sphere of radius $r = R_0$ encompassing the sample, i.e. the left-hand side of

$$\begin{aligned} \Phi(r,\theta,\phi) &= \frac{1}{4\pi}\iiint_Q \left[M_1(\vec{t})(r\sin\theta\cos\phi - t_1) + M_2(\vec{t})(r\sin\theta\sin\phi - t_2)\right. \\ &\left. + M_3(\vec{t})(r\cos\theta - t_3)\right] \\ &\times\frac{d^3t}{\left(r^2-2r\left[(t_1\cos\phi+t_2\sin\phi)\sin\theta+t_3\cos\theta\right]+t_1^2+t_2^2+t_3^2\right)^{3/2}}. \end{aligned} \tag{2}$$

Since Φ is harmonic for $r > R_0$, we can expand it over solid harmonics

$$\Phi(r,\theta,\phi) = \sum_{l=0}^{\infty}\frac{1}{r^{l+1}}\sum_{j=-l}^{l}c_{j,l}S_l^j(\theta,\phi), \quad S_l^j(\theta,\phi) := \begin{cases} P_l^j(\cos\theta)\cos(j\phi), & j \ge 0, \\ P_l^{|j|}(\cos\theta)\sin(|j|\phi), & j < 0, \end{cases}$$

where S_l^j are spherical harmonics and P_l^j are associated Legendre polynomials.

As seen from (2), Φ decays at infinity as $\mathcal{O}\left(1/r^2\right)$ which implies that $c_{0,0} = 0$.

Defining the L^2 inner product on the sphere of radius R in the usual way

$$\langle f, g\rangle_{L^2(\mathbb{S}_R)} = \int_0^{2\pi}\int_0^{\pi} f(r,\theta,\phi)\, g(r,\theta,\phi)\, R^2\sin\theta d\theta d\phi,$$

IOP Conf. Series: Journal of Physics: Conf. Series **904** (2017) 012011 doi:10.1088/1742-6596/904/1/012011

we employ orthogonality of spherical harmonics to obtain

$$\lim_{R\to\infty}\left\langle \Phi, \left(S_1^{-1}, S_1^0, S_1^1\right)^T\right\rangle_{L^2(\mathbb{S}_R)} = \left(-\frac{1}{3}m_2, \frac{1}{3}m_3, -\frac{1}{3}m_1\right)^T = \left(\frac{4\pi}{3}c_{-1,1}, \frac{4\pi}{3}c_{0,1}, \frac{4\pi}{3}c_{1,1}\right)^T,$$

that is,

$$m_1 = -3\left\langle \Phi, S_1^1\right\rangle_{L^2(\mathbb{S}_{R_0})}, \quad m_2 = -3\left\langle \Phi, S_1^{-1}\right\rangle_{L^2(\mathbb{S}_{R_0})}, \quad m_3 = 3\left\langle \Phi, S_1^0\right\rangle_{L^2(\mathbb{S}_{R_0})}.$$

These formulas would exactly solve the moment recovery problem if we had the available data:

(i) on a sphere rather than on a horizontal plane;

(ii) for the potential rather than a component of the field;

(iii) on the complete surface rather than its subset.

To tackle the issue (i), we need to map the data to the sphere.

2.2. Kelvin transformation

Recall that in the complex plane $\mathbb{C}$, the Moebius transform $\frac{z-i}{z+i}$ sends the upper half-plane $\operatorname{Im} z > 0$ onto the unit disk $|z| < 1$ preserving harmonicity. Kelvin transformation is a generalization of this concept to higher dimensions [1]. In particular, we consider a one-parameter ($R_0 > 0$) family of transforms

$$\mathcal{K}[f]\left(\vec{\xi}\right) \equiv f^\star\left(\vec{\xi}\right) = \frac{1}{\left|\vec{\xi}-\vec{s}\right|} f\left(\mathcal{R}\vec{\xi}\right),$$

where $\vec{\xi} := (\xi_1, \xi_2, \xi_3)^T$, $\vec{s} := (0, 0, -R_0)^T$, $e_0 := \sqrt{2R_0(R_0+h)}$, and

$$\mathcal{R}\vec{\xi} := \left(\frac{e_0^2\xi_1}{\xi_1^2+\xi_2^2+(\xi_3+R_0)^2}, \frac{e_0^2\xi_2}{\xi_1^2+\xi_2^2+(\xi_3+R_0)^2}, -R_0 + \frac{e_0^2(\xi_3+R_0)}{\xi_1^2+\xi_2^2+(\xi_3+R_0)^2}\right)^T,$$

which map functions on horizontal plane $x_3 = h$ onto those defined on the sphere of radius R_0 centered at the origin.

Hence, we have:

$$\Delta f(\mathbf{x}, x_3) = 0, \quad x_3 > h \qquad \Longleftrightarrow \qquad \Delta f^\star\left(\vec{\xi}\right) = 0, \quad \left|\vec{\xi}\right| < R_0.$$

Application of this transform to the potential followed by restriction to the sphere $\mathbb{S}_{R_0}$ gives

$$\begin{aligned}\mathcal{K}[\Phi](\theta,\phi) &= \frac{1}{4\pi R_0\sqrt{2(1+\cos\theta)}}\iiint_Q \left[M_1\left(\vec{t}\right)\left(\frac{(R_0+h)\sin\theta\cos\phi}{1+\cos\theta} - t_1\right)\right. \\ &\quad \left. + M_2\left(\vec{t}\right)\left(\frac{(R_0+h)\sin\theta\sin\phi}{1+\cos\theta} - t_2\right) + M_3\left(\vec{t}\right)(h-t_3)\right] \\ &\quad \times \frac{d^3t}{\left[\left(\frac{(R_0+h)\sin\theta\cos\phi}{1+\cos\theta} - t_1\right)^2 + \left(\frac{(R_0+h)\sin\theta\sin\phi}{1+\cos\theta} - t_2\right)^2 + (h-t_3)^2\right]^{3/2}}.\end{aligned}$$

Unfortunately, because of a more complicated angular dependence, the simple link between the net moment components and projections onto the first three spherical harmonics is now broken:

$$\left\langle \mathcal{K}[\Phi], S_1^{-1} \right\rangle_{L^2(\mathbb{S}_{R_0})} \not\propto m_1, \quad \left\langle \mathcal{K}[\Phi], S_1^{1} \right\rangle_{L^2(\mathbb{S}_{R_0})} \not\propto m_2, \quad \left\langle \mathcal{K}[\Phi], S_1^{0} \right\rangle_{L^2(\mathbb{S}_{R_0})} \not\propto m_3.$$

However, it turns out that proportionality in the first two formulas still holds asymptotically for large values of R_0 (as can be expected geometrically), that is, we again have

$$\lim_{R_0\to\infty} \left\langle \mathcal{K}[\Phi], S_1^{-1} \right\rangle_{L^2(\mathbb{S}_{R_0})} \sim m_1, \quad \lim_{R_0\to\infty} \left\langle \mathcal{K}[\Phi], S_1^{1} \right\rangle_{L^2(\mathbb{S}_{R_0})} \sim m_2.$$

Namely, multiscale analysis of integrals (see [4, Sect. 3.3.1]) results in

$$m_j = 6 \lim_{R_0\to\infty} R_0^4 \iint_{\mathbb{R}^2} \Phi(\mathbf{x}, h) \frac{x_j}{\left[x_1^2 + x_2^2 + (R_0 + h)^2\right]^{5/2}} dx_1 dx_2, \quad j = 1, 2.$$

Employing spherical harmonics expansion and a specially derived connection formula for normal derivatives

$$\mathcal{K}[\partial_{x_3}\Phi]\left(\vec{\xi}\right) = -\frac{1}{e_0^2}(R_0 + \xi_3)\left(\mathcal{K}[\Phi]\left(\vec{\xi}\right) + 2R_0\partial_r \mathcal{K}[\Phi]\left(\vec{\xi}\right)\right), \quad \vec{\xi} \in \mathbb{S}_{R_0},$$

we can show that

$$m_j = \frac{2}{\mu_0} \lim_{R_0\to\infty} R_0^5 \iint_{\mathbb{R}^2} B_3(\mathbf{x}, h) \frac{x_j}{\left[x_1^2 + x_2^2 + (R_0 + h)^2\right]^{5/2}} dx_1 dx_2, \quad j = 1, 2, \tag{3}$$

thus overcoming the issue (ii) from the previous subsection.

2.3. Normal component recovery and incomplete data

Interestingly enough, the normal component of the net moment cannot be recovered by asymptotic projection on the spherical harmonic S_1^0 neither from the field nor from the potential. One can, however, proceed another way. We use Poisson representation formula for $\mathcal{K}[\Phi]$ harmonic inside the ball and pass to the limit approaching vertically from inside the south pole $\vec{s}$ which, on the other hand, as an image of infinity under Kelvin transformation, must be proportional to m_3 at the leading order due to the dominance of vertical component term in the formula for B_3. Careful analysis and series of integrations by parts permitting passage to the limit lead to the following result [4, Sect. 3.4]

$$m_3 = -\frac{2}{\mu_0} \lim_{\rho\to\infty} \rho^3 \int_0^{2\pi} B_3(\rho\cos\varphi, \rho\sin\varphi, h)\, d\varphi. \tag{4}$$

Despite its simplicity, this formula is useless in practice for it only involves values of the field in the far region, exactly, where they cannot be measured.

We now focus on (iii), the last of the formulated issues, namely, dealing with incomplete data.

The remedy of situation for the case of normal component comes from application of Gauss theorem by splitting of the integral into a part with available values and its complement

$$0 = \iint_{\mathbb{R}^2} B_3(\mathbf{x}, h)\, dx_1 dx_2 = \iint_{D_A} B_3(\mathbf{x}, h)\, dx_1 dx_2 + \iint_{\mathbb{R}^2 \setminus D_A} B_3(\mathbf{x}, h)\, dx_1 dx_2.$$

Assuming the measurement area D_A is large enough, we use, in the second term on the right, a far-distance asymptotic expansion of the field (1) whose integral is proportional to m_3. This immediately gives

$$m_3 = \frac{2}{\mu_0 A} \iint_{D_A} B_3(\mathbf{x}, h)\, dx_1 dx_2 + \mathcal{O}\left(\frac{1}{A^2}\right), \tag{5}$$

which provides a practical replacement of the formula (4).

Applying the same splitting and asymptotic extension strategy to (3), we arrive at

$$m_j = \frac{2}{\mu_0} \iint_{D_A} B_3(\mathbf{x}, h)\, x_j dx_1 dx_2 + \mathcal{O}\left(\frac{1}{A}\right), \quad j = 1, 2. \tag{6}$$

3. Fourier analysis

Define Fourier transform as $\mathcal{F}[f](\mathbf{k}) = \hat{f}(\mathbf{k}) = \iint_{\mathbb{R}^2} f(\mathbf{x}) e^{2\pi i \mathbf{k}\cdot\mathbf{x}} dx_1 dx_2$, $\mathbf{k} := (k_1, k_2)^T$, and observe that $\mathcal{F}\left[\left(x_1^2 + x_2^2 + H^2\right)^{-3/2}\right](\mathbf{k}) = \frac{2\pi}{H} e^{-2\pi H|\mathbf{k}|}$ for $H > 0$.

Then, representing the field as

$$\begin{aligned} B_3(\mathbf{x}, h) &= -\frac{\mu_0}{4\pi} \iiint_Q \left[(h - t_3)\left(M_1(\mathbf{t}, t_3) \left.\frac{\partial}{\partial x_1}\right|_{x_3=h} + M_2(\mathbf{t}, t_3) \left.\frac{\partial}{\partial x_2}\right|_{x_3=h}\right)\right. \\ &\quad \left. + M_3(\mathbf{t}, t_3) \left.\frac{\partial}{\partial x_3}\right|_{x_3=h} (x_3 - t_3)\right] \left(|\mathbf{x} - \mathbf{t}|^2 + (x_3 - t_3)^2\right)^{-\frac{3}{2}} d^3 t, \end{aligned}$$

and denoting Q_3 the vertical projection of the magnetization support set Q, we take advantage of convolution structure of the integral operator and arrive at

$$\hat{B}_3(\mathbf{k}, h) = \pi\mu_0 \int_{Q_3} e^{-2\pi(h-t_3)|\mathbf{k}|} \left[ik_1 \hat{M}_1(\mathbf{k}, t_3) + ik_2 \hat{M}_2(\mathbf{k}, t_3) + |\mathbf{k}|\, \hat{M}_3(\mathbf{k}, t_3)\right] dt_3. \tag{7}$$

Note that $\vec{m} = \int_{Q_3} \hat{\vec{M}}(\mathbf{0}, t_3)\, dt_3$, and $\hat{\vec{M}}(\mathbf{k}, t_3)$ is smooth in $\mathbf{k}$ due to compact support of $\vec{M}$ (by Paley-Wiener theorem). Hence to extract the net moment information, we expand about $\mathbf{k} = \mathbf{0}$ all the right-hand side terms as well as the first integral in $\hat{B}_3(\mathbf{k}, h) = \left(\iint_{D_A} + \iint_{\mathbb{R}^2 \setminus D_A}\right) e^{2\pi i \mathbf{k}\cdot\mathbf{x}} B_3(\mathbf{x}, h)\, dx_1 dx_2$ and, in the second one, we use continuation of B_3 beyond the measurement area according to its asymptotics as discussed before. Matching terms of different smallness in $\mathbf{k}$, and employing asymptotic expansions of some special functions (Bessel, Struve, sine integral), we obtain a set of relations connecting field integrals against monomials with algebraic magnetization moments of different order (see [4, Sect. 3.6] for more details). In particular, taking the imaginary part of (7) and fixing either $k_1 = 0$ or $k_2 = 0$ we are led to formulas (6) whereas working with real part of (7) results in (5). More sophisticated combinations of the obtained relations yield asymptotic formulas which are superior to (6)

$$m_j = \frac{2}{\mu_0} \iint_{D_A} \left(1 + \frac{4x_j^2}{3A^2}\right) x_j B_3(\mathbf{x}, h)\, dx_1 dx_2 + \mathcal{O}\left(\frac{1}{A^2}\right), \quad j = 1, 2. \tag{8}$$

4. Numerical illustrations

We demonstrate results by performing numerical simulation on a synthetic example of few magnetic dipoles (i.e. point-supported magnetization sources) with positions $\vec{x}^{(1)} = (3.5, 3.0, 1.0)^T\, 10^{-5}$ m, $\vec{x}^{(2)} = (0.0, 0.0, 7.0)^T\, 10^{-5}$ m, $\vec{x}^{(3)} = (4.0, -5.5, 11.5)^T\, 10^{-5}$ m, $\vec{x}^{(4)} = (-4.0, 5.5, 2.5)^T\, 10^{-5}$ m and magnetic moments $\vec{m}^{(1)} = (4.5, 3.5, 1.0)^T\, 10^{-12}$ A·m^2, $\vec{m}^{(2)} =$

$(2.5, 4.5, 0.5)^T\, 10^{-12}$ A·m^2, $\vec{m}^{(3)} = (-3.0, 2.0, 2.5)^T\, 10^{-12}$ A·m^2, $\vec{m}^{(4)} = (-1.0, 2.0, 1.5)^T\, 10^{-12}$ A·m^2 producing a magnetic field measured at $x_3 = h = 2.5 \cdot 10^{-4}$ m.

Due to their asymptotic nature, the quality of the obtained formulas depends on the size of the measurement area: the bigger the area, the better the accuracy. We plot the estimates of the net moment components for different values of measurement disk radius A comparing the first and the second order accuracy formulas (6), (8) for the tangential components m_1, m_2, and illustrating the estimate for the normal component m_3 given by formula (5) which is already of the second order. Figure 1 contains results for the noiseless field whereas Figure 2 shows the same estimate for the field contaminated with Gaussian white noise of SNR=20 dB. We observe persistence of the estimates for m_1 and m_2 with respect to the noise while the estimate for m_3 has to be corrected to remedy amplification of noise due to factor A (in particular, we use 11-point linear regression to suppress asymptotically linear growth in A excluding from consideration the region of lower values of A). It is also clear that magnetic components of smaller magnitudes are more deteriorated by the noise.

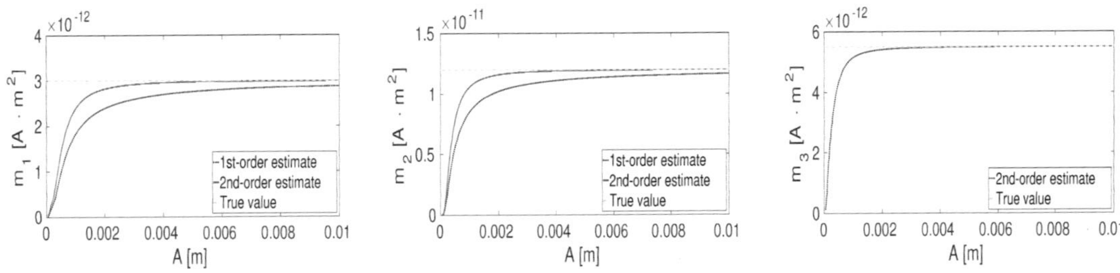

Figure 1. Estimates for m_1 (left), m_2 (center), m_3 (right) versus A, without noise.

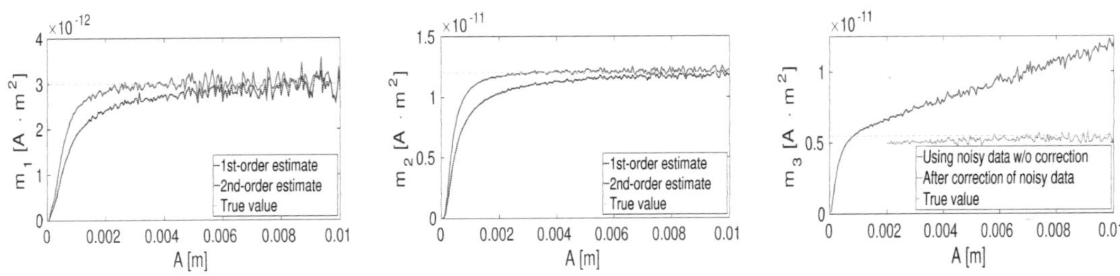

Figure 2. Estimates for m_1 (left), m_2 (center), m_3 (right) versus A, with noise of SNR=20.

5. Conclusion

The source feature (net moment vector) recovery issue in the context of inverse magnetization problem has been considered. We have devised two strategies of solving the problem which can be extended to more general frameworks. The second-order asymptotic formulas for estimating magnetization net moment components from the partially available data (5), (8) constitute the main practical outcome of the work. Residue terms in the obtained formulas can potentially be computed with arbitrary high precision and, by means of combination of measurements from different size areas, can be reduced to higher order of smallness. Some combinations of higher-order algebraic moments can also be easily extracted from the field data. The formulas extend to other geometries with the only difference being in numerical factors; in particular, rectangular geometry has been considered as well, and the corresponding results will be published elsewhere.

IOP Conf. Series: Journal of Physics: Conf. Series **904** (2017) 012011 doi:10.1088/1742-6596/904/1/012011

Acknowledgements
The authors would like to thank Doug Hardin (Vanderbilt University) and Sylvain Chevillard (INRIA Sophia Antipolis) for fruitful discussions.

References
[1] Axler S et al 2001 *Harmonic Function Theory* (New York: Springer-Verlag)
[2] Baratchart L, Hardin D P, Lima E A, Saff E B, Weiss B P 2013 Characterizing kernels of operators related to thin-plate magnetizations via generalizations of Hodge decompositions *Inverse Probl.* **29**(1)
[3] Lima E A, Weiss B P, Baratchart L, Hardin D P, Saff E B 2013 Fast inversion of magnetic field maps of unidirectional planar geological magnetization *J. Geophys. Res.: Solid Earth* **118**(6), 2723-52
[4] Ponomarev D 2016 *Some inverse problems with partial data*, PhD dissertation (University of Nice)

IOP Conf. Series: Journal of Physics: Conf. Series **904** (2017) 012012 doi:10.1088/1742-6596/904/1/012012

Remarks on a financial inverse problem by means of Monte Carlo Methods

Salvatore Cuomo, Vittorio Di Somma, Federica Sica
Universitá degli Studi di Napoli Federico II, Strada Vicinale Cupa Cinthia, 21, Naples, Italy
E-mail: {salvatore.cuomo@unina.it, vittorio.disomma}@unina.it

Abstract. Estimating the price of a barrier option is a typical inverse problem. In this paper we present a numerical and statistical framework for a market with risk-free interest rate and a risk asset, described by a Geometric Brownian Motion (GBM). After approximating the risk asset with a numerical method, we find the final option price by following an approach based on sequential Monte Carlo methods. All theoretical results are applied to the case of an option whose underlying is a real stock.

1. Introduction

Recently, the growth of financial markets, especially due to the diffusion of financial instruments as derivatives, has forced analysts to create efficient mathematical models to estimate risk assets trends. Since a risk asset can be identified, formally, with a Stochastic process respect to a fixed probability space, this problem is generally an *inverse problem* and involves the solution of differential equations and the estimation of such parameters.

Here we deal with European double barrier options, a particular kind of *exotic options*, which differ from *vanilla options* for the complexity of their financial structure. More in detail, this concerns the definition of the underlying or payoff function. A double barrier option is characterized by the presence of two numerical values L, U, $(L < U)$, called *barriers*. Moreover, two cases are possible: (i) a *Knock-In barrier option* that comes to life when the value of the underlying arrives at one of the barrier values; (ii) a *Knock-out Barrier Option* that concludes its life when the value of the underlying arrives at one of the barrier values.

Generally, a barrier option is less expensive than the vanilla option with the same underlying, strike price and maturity. Conversely, respect to the last one, it has a higher riskiness degree: the former property implies that traders are attracted more by a barrier option, the last property implies that a barrier option is suitable to a vanilla one when the underlying value has a regular trend. The main reference for options , and in general for derivatives, is in [8]; in particular, for the problem of option pricing we also refer to [2] and [3].

From a mathematical point of view, the problem of barrier option pricing is addressed by Monte Carlo methods and in this paper we adopt a *sequential Monte Carlo approach*: for a complete description of this topic see [9], [4] and [5], for the implementation issues you can also see [6] and [7]. Finally, we propose a stochastic model to describe the market and a numerical procedure to give a barrier option price.

The paper is organized in the following way: in the section 2 we give some mathematical preliminaries; the section 3 is devoted to the pricing procedure and a real real case study; in the last section we draw the conclusions.

2. Mathematical and Financial preliminaries

The option pricing can be interpreted as a classical example of *inverse problem.* From a formal point of view, we have two stochastic processes Y_t and X_t, where Y_t represents a phenomenon and X_t its stochastic casual factors, and we suppose that a functional dependence between them, indicated with F_t, exists:

$$Y_t = F_t(X_t, \theta_t),$$

where θ_t is a time-depending vector. An inverse problem aims at the estimation of F_t and θ_t, given a set of observations of Y_t.
In our context, the stochastic process Y_t is identified as the option price, the stochastic process as the underlying, the parameter vector is the risk-free interest rate and the volatility of the underlying. In this paper we deal with the determination of the function F_t and, here, we assume that in our market the interest rate and the volatility are known.

In the following we fix a time interval $[0;T]$, a Brownian motion W_t and a probability space $(\Omega;\mathbb{F}_t;\mathbb{P})$ for all stochastic processes; we denote the constant risk-free interest rate with r. The lognormal risk asset S_t with constant volatility σ and drift μ solves the following stochastic differential equation:

$$dS(t) = \mu S(t)dt + \sigma S(t)dW_t. \tag{1}$$

The drift parameter is the difference between the interest rate r and a constant dividend q. The no arbitrage vanilla option price P_0^{bs} at the instant 0 is given by the *Black-Scholes formula* (the Black-Scholes model is described in details in [1]):

$$P_0^{bs} = \begin{cases} S_0\Phi(d_1) - Ke^{-rT}\Phi(d_2) & \text{if} \quad a) \\ Ke^{-rT}\Phi(-d_2) - S_0\Phi(-d_1) & \text{if} \quad b) \end{cases} \tag{2}$$

where $a)$ refers to a **Call**, which gives the traders the right to buy a quantity of the underlying at the maturity by paying an amount K (named *strike price*), while $b)$ refers to a **Put**, which gives the right to sell a quantity of the underlying at the maturity by receiving an amount K (as in the case of a Call, it is called *strike price*). Moreover, in the previous formula, it holds:

$$\begin{aligned} d_1 &= \frac{\log\left(\frac{S_0}{K}\right) + \left(r + \frac{\sigma^2}{2}\right)T}{\sigma\sqrt{T}} \\ d_2 &= d_1 - \sigma\sqrt{T} \quad \Phi(d) = \frac{1}{\sqrt{2\pi}}\int_{-\infty}^{d} e^{-\frac{x^2}{2}}\, dx. \end{aligned}$$

The pay-off function H of an option is defined in the following way:

$$H(S_t) = \begin{cases} \max\{S_T - K, 0\} & \text{if Call} \\ \max\{K - S_T, 0\} & \text{if Put} \end{cases}$$

In the following we indicate the expected value of a random variable X with $E[X]$ and the indicator function of an interval $]a;b[$ with $1_{]L;U[}(x)$.
The prices of a knock-in and knock-out barrier option at 0, with underlying S_t and pay-off function $H(S_t)$, are expressed in the following way:

$$\begin{cases} P_0^{out} = e^{-rT} E\left(H(S_T)\prod_{t\in[0;T]} 1_{]L;U[}(S_t)\right) \\ P_0^{in} = P_0^{bs} - P_0^{out}. \end{cases} \tag{3}$$

We deal with the resolution of the described problems in the next section.

3. Numerical pricing procedure

We describe a numerical procedure to apply the model defined in the previous section. We decompose the time interval $[0;T]$ with N equidistant points $0 = t_0 < ... < t_{N-1} = T$ and we indicate the value of the underlying at the instant t_n and the discretization step respectively with the symbol S_n and h; for every n, the variables $H(S_N)$ and $1_{]L;U[}(S_n)$ are uncorrelated, the following approximations of the knock-out and knock-in barrier option prices are valid:

$$\begin{cases} P_0^{out} \sim e^{-rT} E[H(S_{N-1})] \prod_{n=0}^{N-1} E\left[1_{]L;U[}(S_n)\right] \\ P_0^{in} \sim P_0^{bs} - P_0^{out}. \end{cases} \tag{4}$$

At every step n, our numerical procedure consists in approximating the value S_n of the previous system by the mean of M particles and in estimating the price values in (4) by a Monte Carlo approach. More in detail, our schema is composed of the following steps: i) generation of a number M of the particles $S_n^{(m)}$, with $1 \leq m \leq M$ and $0 \leq n \leq N-1$; ii) rejection of the underlying values (the particles) with the lowest probability to stay in the interval $]L, U[$; iii) determination of the values in (4).

We start to analyze the point i). We set $X_t := logS_t$ and we transform the system (1) into the following one:

$$\frac{dX_t}{dt} = \mu + \sigma \frac{dW_t}{dt}. \tag{5}$$

The numerical solutions $X_n^{(m)}$ of this equation, which represent the logarithms of the particles $S_n^{(m)}$, are found by applying a *first-order numerical method*, as it follows:

$$X_{n+1}^{(m)} = \mu_n^m + \sqrt{h}\sigma z_{n+1}^{(m)} \quad z_{n+1}^{(m)} \sim N(0;1) \quad \mu_n^m = X_n^{(m)} + hr$$

As it concerns the point ii), we (re)sample these just obtained particles. This procedure is made by two suitable functions g and G defined in the following way:

$$g(X_{n+1}^{(m)}; X_n^{(m)}) \quad := \quad \mathbb{P}(X_{n+1}^{(m)} \in]\ln L; \ln U[\mid X_n^{(m)}) = \frac{1}{\sigma\sqrt{2h\pi}} \int_{\log L}^{\log U} e^{-\frac{(x-\mu_n^m)^2}{2h^2\sigma^2}}\, dx. \tag{6}$$

$$G(X_{n+1}^{(m)}, X_n^{(m)}) \quad = \quad \begin{cases} g(X_{n+1}^{(m)}; X_n^{(m)}), & \text{if} \quad X_{n+1}^{(m)} \in]\log L; \log U[\\ 0, & \text{otherwise.} \end{cases} \tag{7}$$

We briefly explain the re-sampling technique. We only choose the particles with the highest values of the function G. At every step, we summarize the values of the function G until this sum does not become greater than a drawn uniform number u_k, with k an integer, and we indicate the number of the addends with s_k; at this point, we set all the previous particles equal to s_k-th particle. For every time step, this procedure stops when the sum of these indexes s_k is equal to M, in order to have always a number M of particles.

For the point iii), estimators for the knock-out and knock-in price P_{out} and P_{in} have been determined by making use of the **law of large numbers** :

$$\begin{cases} P_0^{out} = e^{-rT}\left(\frac{1}{M}\sum_{m=1}^{M} H\left(S_{N-1}^{(m)}\right)\right)\prod_{n=0}^{N-1}\frac{1}{M}\sum_{m=1}^{M} 1_{]L;U[}\left(S_n^{(m)}\right) \\ P_0^{in} = P_0^{bs} - P_0^{out}. \end{cases} \quad (8)$$

The algorithm which implements our schema is summarized in the following.
The input variables are: i) the upper bound T of the time interval; ii) the number N of time steps; iii) the number M of particles at every time step; iv) the barriers L and U; v) the strike price K; vi) the risk-free interest rate r; vii) the volatility σ; vii) the current value of the underlying S_0; viii) the theoretical Black-Scholes vanilla price P_0^{bs}.
The output variables are: i) the knock-out barrier option price P_0^{out}; ii) the knock-in barrier option price P_0^{in}.
We conclude the section with the application of the proposed framework to a real case. More in detail, we have studied the pricing problem of a knock in down put with a *single barrier* B (i.e. in our model we have set $L = -\infty$ and $U = B$), with **Volkswagen company** stocks as underlying and 6M EURIBOR as the risk-free interest rate. We have applied our procedure and compared the values and the ones of a standard Monte Carlo method (as presented in [10]) with the real price $P = 0.36$€ for different values of N and M between 10 and 100. In the following we list all the values of the input variables:

$$(r, q, S_0, B, K, \sigma, T, P_0^{bs}) = (0.0056, 0.00005, 75, 100, 0.09, 1 \text{ year}, 78.46\text{€}).$$

We have inserted the results into the following table, constructed in the following way: i) the first and the second column contain respectively the values of N and M; ii) the third, the fourth and the fifth columns contain respectively the estimations of the price by the algorithm (3)(indicated with *Sequential*), the corresponding absolute error values (indicated with *Error Seq*) and the average time to complete a simulation (indicated with *Time Seq*); iii) the sixth, the seventh and the eighth columns contain respectively the approximations of the price by the standard Monte Carlo (indicated with *Standard* method), the corresponding absolute error values (indicated with *Error Stan*) and he average time to complete a simulation (indicated with *Time Stan*). All the numerical values have been expressed in Euro.

N	M	Sequential	Error Seq	Time Seq	Standard	Error Stan	Time Stan
200	200	0,68	0,31	1,11	0,99	0,63	0,02
200	150	0,44	0,08	0,83	0,79	0,43	0,02
150	150	0,60	0,24	0,64	0,33	0,03	0,13
100	150	0,42	0,07	0,47	1,06	0,70	0,03
100	100	0,60	0.24	0,43	0,69	0,33	0,01
80	80	0,55	0,19	0,18	0,73	0,37	0,01
80	20	0,35	0,01	0,06	0,51	0,15	0,003
60	50	0,16	0,26	0,09	0,77	0,41	0,004
60	30	0,37	0,01	0,06	1,43	1,07	0,003
40	50	0,04	0,33	0,07	1,23	0,87	0,003
20	50	0,43	0,07	0,07	0,34	0,02	0,01
20	20	0,84	0,48	0,03	0,85	0,49	0,003
10	20	0,78	0,42	0,01	1,49	1,13	0,003

Table 1. Simulations of barrier option prices for different values of N and M.

IOP Conf. Series: Journal of Physics: Conf. Series **904** (2017) 012012 doi:10.1088/1742-6596/904/1/012012

Algorithm 1 Algorithm for Barrier Option pricing

Require: $T, M, N, L, U, K, r, \sigma, S_0, P_0^{bs}$
Ensure: $P_0^{(out)}, P_0^{(in)}$

$$X_1^{(m)} = \log S_0^{(m)}; \quad h = \frac{T}{N} \quad m = 1, ..., M. \quad //\text{Inizialization}$$

for $n = 1, N-1$ **do**
  for $m = 1, M$ **do**

$$\begin{aligned} X_{n+1}^{(m)} &= X_n^{(m)} + hr + \sqrt{h}\sigma z_{n+1}^{(m)}; \quad z_{n+1}^{(m)} \sim N(0;1). \\ \mu_n^m &= X_n^{(m)} + hr. \quad //\text{Generation of the values} \quad X_n^{(m)} \\ g(X_{n+1}^{(m)}; X_n^{(m)}) &= \frac{1}{\sigma\sqrt{2h\pi}} \int_{\ln L}^{\ln U} e^{-\frac{(x-\mu_n^m)^2}{2(h\sigma)^2}} dx. \end{aligned}$$

  if $X_{n+1}^{(m)} \in]L; U[$ **then**

$$G^1(X_{n+1}^{(m)}; X_n^{(m)}) = g(X_{n+1}^{(m)}; X_n^{(m)}) \quad G(X_{n+1}^{(m)}; X_n^{(m)}) = \frac{G^1(X_{n+1}^{(m)})}{\sum_{k=1}^M G^1(X_{n+1}^{(k)}; X_n^{(k)})}.$$

  else $G(X_{n+1}^{(m)}; X_n^{(m)}) = 0.$ //Computing of the functions g and G
  end if
  end for
  for $m = 1, M$ **do** $(G = 0 \quad j = 1 \quad u^{(m)} \sim Uniform(0,1))$
  while $G < u^{(m)}$ and $j \leq M$ **do** //Re-sampling of the particles.

$$G = \sum_{k=1}^{j} G(X_{n+1}^{(k)}); X_n^{(k)}) \quad j \leftarrow j+1.$$

  end while

$$X_{n+1}^{(k)} = X_{n+1}^{(j)}; \quad G(X_{n+1}^{(k)}; X_n^{(k)}) = G(X_{n+1}^{(j)}; X_n^{(j)}) \quad k = 1, ..., j-1.$$

  end for
end for

$$\begin{aligned} S_n^{(m)} &= exp(X_n^{(m)}); \quad P_0^{out} = e^{-rT}\left(\frac{1}{M}\sum_{m=1}^{M} H\left(S_{N-1}^{(m)}\right)\right) \prod_{n=0}^{N-1} \frac{1}{M} \sum_{m=1}^{M} 1_{]L;U[}\left(S_n^{(m)}\right) \\ P_0^{in} &= P_0^{bs} - P_0^{out}. \quad //\text{Option pricing} \end{aligned}$$

Our results suggest that:

- for high values of N and M both the sequential and the standard Monte Carlo methods have similar results in mean;
- for low values of N and M the sequential Monte Carlo method gives better results than a standard Monte Carlo procedure.

In conclusion, our method is more suitable than a standard Monte Carlo one respect to low values of N and M: this is due in particular to three factors: i) the discretization of the underlying; ii) the re-sampling technique; iii) the chosen price estimator.

4. Conclusions

The procedure of pricing the financial options is an inverse problem that involves several numerical and statistical issues, as for example the Monte Carlo methods and differential equation methods. We have proposed a numerical procedure to estimate barrier options under the assumptions of the Black-Scholes model: (i) a risk asset described by a Geometric Brownian motion; (ii) arbitrages are avoided; (iii) short sellings are allowed.
In a future work, we will extend the proposed approach to the case of a non constant interest rate and volatility and we will develop a model for their estimation.

References

[1] Black F and Scholes M 1973 The Pricing of Options and Corporate Liabilities *J. Polit. Econ.* **81** 637-654.
[2] Cuomo S, De Michele P, Di Somma V and Galletti A 2016 A Stochastic Method for Financial IoT data *Procedia Comput Sci* (Amsterdam: Elsevier) **98** 491-496.
[3] Cuomo S, De Michele P, Di Somma V and Ponti G 2016 A Probabilistic Approach for Financial IoT data *Midas Garda Proc.*
[4] Calvetti D, Kaipio J P and Somersalo E 2014 Inverse Problems in the Bayesian Framework *Inverse Problems* **30** 11.
[5] Calvetti D, Cuomo S, Pragliola M, Somersalo E, Toraldo G 2016 Computational Issues and Numerical Experiments for Linear Multistep Method Particle Filtering *AIP Conf. Proc.* **1776** (College Park, Maryland: American Institute of Physics Inc) 040009.
[6] Sen D, Jasra A and Zhou Y 2016 Some contributions to sequential Monte Carlo methods for option pricing *JSCS* (UK: Taylor and Francis Group).
[7] Shevchenko P V and Del Moral P 2016 Valuation of Barrier Options using Sequential Monte Carlo *Journal of Computational Finance* **4** .
[8] Hull J 2009 *Options, Futures,and Other Derivatives* (London:Pearson).
[9] Glasserman P 2004 *Monte Carlo Methods in Financial Engineering* vol 53 of Applications of Mathematics (New York: Springer-Verlag).
[10] Pascucci A 2007 *Calcolo Stocastico per la Finanza* (Milan:Springer).

IOP Conf. Series: Journal of Physics: Conf. Series **904** (2017) 012013 doi:10.1088/1742-6596/904/1/012013

A fast gradient projection method for 3D image reconstruction from limited tomographic data

V.L. Coli[1], E. Loli Piccolomini[2], E. Morotti[3] and L. Zanni[1]

[1] Department of Physics, Informatics and Mathematics, University of Modena and Reggio Emilia

[2] Department of Mathematics, University of Bologna

[3] Department of Mathematics, University of Padova

E-mail: elena.loli@unibo.it

Abstract. We consider in this paper the problem of reconstructing 3D Computed Tomography images from limited data. The problem is modeled as a nonnegatively constrained minimization problem of very large size. In order to obtain an acceptable image in short time, we propose a scaled gradient projection method, accelerated by exploiting a suitable scaling matrix and efficient rules for the choice of the step-length. In particular, we select the step-length either by alternating Barzilai-Borwein rules or by exploiting a limited number of back gradients for approximating second-order information. Numerical results on a 3D Shepp-Logan phantom are presented and discussed.

1. Introduction

3D Computed Tomography (CT) is a well known technique used in different areas, such as medicine, industry or art, to visually represent the interior of a 3D object. The image processing consists in solving an inverse problem, where the data are represented by the so called projections that are emitted by a source and are partially absorbed by the object before being collected on a 2D detector. The source moves around the object through a fixed number of positions (angles). The 3D image visualization consists in a set of 2D images stacked along the third direction. Different techniques exist for the image reconstruction and they are satisfactory when a large set of data is available. At present, the case of limited data is of great interest especially in medical CT, where it is desirable to have less radiations on the patient and to shorten the exam for different reasons. The sub-sampling of the data usually derives from a limited number of source angles [1, 2].

The solution of the image reconstruction problem from limited data can be performed by exploiting priors on the solution, such as the gradient sparsity. The problem is in this case formulated as a constrained or unconstrained minimization problem and solved by iterative algorithms [3]. Some different approaches have been proposed for the solution of the optimization problem, such as first-order methods with the acceleration proposed by Nesterov [4], a soft-thresholding filtering approach [5] or a fixed point method [6]. Since the problem has huge dimensions, of millions of variables, the iterative algorithms are appealing in the real systems only if they obtain a good image reconstruction in very few iterations, i.e. if the convergence speed is high in the first iterations. In this paper we present a Scaled Gradient Projection (SGP) method [7] with such a property. The proposed approach exploits both special scaled gradient directions and suitable step-length selection rules. In particular, we consider two different strategies for the choice of the step-length: the alternating Barzilai-Borwein rule proposed in [8] and a technique recently developed in [9] by generalizing to constrained problems the idea given in [10].

Numerical experiments on a 3D Shepp Logan phantom with limited data affected by Poisson noise are carried out for evaluating the effectiveness of the proposed scaling strategy and step-length selections.

The paper is organized as follows. In Section 2 we describe the optimization problem modeling the image reconstruction; in Section 3 we describe the proposed SGP algorithm; in Section 4 we present some numerical results and finally in Section 5 we report our conclusions.

2. The numerical model

The CT image reconstruction model arising from the discretization of the Lambert-Beer law [11] in the monochromatic case is:

$$Ax = b \tag{1}$$

where $b \in R^{N_p}$ is the vector of recorded projections affected by noise, $x \in R^{N_v}$ represents the discretized 3D object and $A \in R^{N_p \times N_v}$ is the matrix describing the system geometry obtained by using the Siddon algorithm, based on geometrical ray-tracing. The forward projection process of an object and the back projection of a data vector are described by matrix-vector products involving A and A^T, respectively. In this work we suppose the Poisson noise deriving from the particles dynamic dominating with respect to the Gaussian noise due to the digital encoding. In the case of view angle undersampling, the number of data is less than the number of voxels of the discretized object ($N_p < N_v$), hence the linear system (1) might have infinite solutions. Moreover, the problem is ill-conditioned and some of the solutions of (1) are dominated by noise. Hence we reformulate the problem as a constrained regularized minimization problem of the form:

$$\min_{x \geq 0} f(x) = J(x) + \lambda R(x) \tag{2}$$

where $J(x)$ is the fit-to-data function, $R(x)$ is the regularization function and λ is the regularization parameter. In the case of Poisson noise a suitable function $J(x)$ is the Kullback-Leibler divergence:

$$J(x) = \sum_{i=1}^{N_p} \left(\sum_{j=1}^{N_v} A_{ij}x_j + bg - b_i - b_i log \frac{\sum_{j=1}^{N_v} A_{ij}x_j + bg}{b_i} \right) \tag{3}$$

where $bg > 0$ is the background value. For what concerns the regularization function $R(x)$, the majority of CT reconstruction algorithms exploit the sparsity in the gradient magnitude image, using regularization functions involving ℓ_1 seminorms, such as the Total Variation (TV) function [1, 2, 12]. The reason is that since tomographic images are rather uniform inside the organs, while they have fast variations in the borders of the organs, the TV function seems very suitable for their approximation. In this paper we consider as regularization function $R(x)$ the TV function in its discrete smoothed form, denoted by $TV_\beta(x)$:

$$R(x) = TV_\beta(x) = \sum_{i=1}^{N_v} (\|\nabla x_i\|_2^2 + \beta^2)^{1/2} \tag{4}$$

where β is a positive small parameter. The constraint $x \geq 0$ is due to the physical meaning of the solution x representing the value of the attenuation coefficient, that is nonnegative in each voxel of the discretized object. Because of the huge size of the problem (N_p is of order 10^6, N_v is of order 10^9), in real applications the matrix A is never stored in the computer memory, but it is re-computed whenever a matrix-vector product (involving A or A^T) is required. Since in this paper we deal with small size test problems, we store the nonzero elements of A.

3. The SGP algorithm for CT image reconstruction

Due to the possible huge size of the matrix A and the very simple constraints of problem (2), first-order algorithms exploiting only the gradient of $f(x)$ and the projection onto the feasible region are very appealing approaches. Among these algorithms, the gradient projection methods are the most popular and, thanks to recent ideas for accelerating their convergence rate, they have given rise to very effective image reconstruction algorithms in many different areas. For these reasons, in the sequel we recall the bases of the main acceleration strategies for gradient methods and we discuss how they can be exploited for designing an efficient method for our reconstruction problem. The basic ideas to accelerate gradient projection methods concern essentially the choice of the gradient-based direction and the selection of the step-length parameter controlling the movement along this direction. In order to define effective gradient-based directions, very promising results have been obtained by scaling the negative gradient by means of suitable positive definite diagonal matrices [13, 7, 14], while, for what concerns the step-length selections, adaptive rules able to improve the convergence rate of standard gradient methods have been derived following the pioneering work of Barzilai-Borwein (BB) [15, 16, 8, 10]. A general scheme able to include both the above strategies is given by the SGP algorithm [7] described in Table (1).

Table 1: Algorithm SGP (Scaled Gradient Projection).

Initialize: $x^{(0)} \geq 0,\ \delta, \sigma \in (0,1),\ 0 < \alpha_{min} \leq \alpha_{max},\ \alpha_0 \in [\alpha_{min}, \alpha_{max}],\ D_0 \in \mathcal{D}_{\rho_0}$;
for $k = 0, 1, \ldots$
  $d^{(k)} = P_+\left(x^{(k)} - \alpha_k D_k \nabla f(x^{(k)})\right) - x^{(k)}$; (scaled gradient projection step)
  $\eta_k = 1$;
  while $f(x^{(k)} + \eta_k d^{(k)}) > f(x^{(k)}) + \sigma \eta_k \nabla f(x^{(k)})^T d^{(k)}$
    $\eta_k = \delta \eta_k$; (backtracking step)
  end
  $x^{(k+1)} = x^{(k)} + \eta_k d^{(k)}$;
  define the diagonal scaling matrix $D_{k+1} \in \mathcal{D}_{\rho_{k+1}}$; (scaling updating rule)
  define the step–length $\alpha_{k+1} \in [\alpha_{min}, \alpha_{max}]$; (step–length updating rule)
end

3.1. *The scaling strategy*

In the SGP algorithm, for an N_v dimensional problem, the notation $\mathcal{D}_{\rho_k}$ is used to define the set of diagonal matrices with entries $d_{j,j}^{(k)}$, $j = 1, \ldots, N_v$, such that $\frac{1}{\rho_k} \leq d_{j,j}^{(k)} \leq \rho_k$, with $\rho_k > 1$, and $P_+(z)$ denotes the euclidean projection of the vector $z \in R^{N_v}$ onto the nonnegative orthant. At each SGP iteration, starting from the current iteration $x^{(k)}$, a step along the scaled gradient direction $-D_k \nabla f(x^{(k)})$ is performed by exploiting the parameter $\alpha_k > 0$, defining the length of the step. The resulting vector is projected onto the feasible region to generate the descent direction $d^{(k)}$ and the linesearch procedure ensures a sufficient descent of f along $d^{(k)}$. Finally, the scaling matrix and the step-length need to be updated for the next iteration. Under suitable assumptions on ρ_k, that is $\rho_k^2 = 1 + \zeta_k$, $\zeta_k \geq 0$, $\sum_{k=0}^{\infty} \zeta_k < \infty$, the sequence generated by SGP converges to the solution of our strictly convex minimization problem [17].

For updating the scaling matrix at each iteration, the strategy introduced in [13], based on a splitting of the objective gradient, has shown promising results in terms of convergence acceleration in many applications of scaled gradient methods. This updating strategy consists in defining the entries $d_{j,j}^{(k+1)}$, $j = 1, \ldots, N_v$, of the diagonal matrix D_{k+1} as

$$d_{j,j}^{(k+1)} = \min\left(\rho_{k+1}, \max\left(\frac{1}{\rho_{k+1}}, \frac{x_j^{(k+1)}}{V_j^J(x^{(k+1)}) + \lambda V_j^{TV}(x^{(k+1)})}\right)\right),$$

where the vectors $V^J(x)$ and $V^{TV}(x)$ are such that $\nabla J(x) = V^J(x) - U^J(x)$, $V^J(x) > 0$, $U^J(x) \geq 0$, and $\nabla TV(x) = V^{TV}(x) - U^{TV}(x)$, $V^{TV}(x) > 0$, $U^{TV}(x) \geq 0$. Taking into account the structure of $\nabla J(x)$, a natural choice for $V^J(x)$ is $V^J(x) = A^T\mathbf{1}$, while we set $V^{TV}(x)$ as proposed in [14]. The parameter ρ_{k+1} is chosen as $\rho_{k+1} = \sqrt{1 + 10^{15}/(k+1)^{2.1}}$.

3.2. Two effective choices of the step-length

Aim of this paper is investigating two strategies for an effective choice of the SGP step-length: the adaptive alternation of the two BB rules suggested in [7], and the updating strategy proposed in [9], based on the use of a limited number of back gradients for capturing second-order information. Given the diagonal matrix D_{k+1}, the scaled version of the BB rules provides the values $\alpha_{k+1}^{BB1} = \frac{\|\bar{s}_k\|^2}{\bar{s}_k^T y_k}$, $\alpha_{k+1}^{BB2} = \frac{s_k^T \bar{y}_k}{\|\bar{y}_k\|^2}$, where $s_k = (x^{(k+1)} - x^{(k)})$, $\bar{s}_k = D_{k+1}^{-1} s_k$, $y_k = (\nabla f(x^{(k+1)}) - \nabla f(x^{(k)}))$, $\bar{y}_k = D_{k+1} y_k$; the adaptive alternation we consider is stated as follows:

$$\alpha_{k+1}^{\mathrm{ABB}} = \begin{cases} \min\left\{\alpha_j^{\mathrm{BB2}} : j = \max\{1, k+1-m_\alpha\}, \ldots, k+1\right\}, \; \tau = 0.9\,\tau, & \text{if } \dfrac{\alpha_{k+1}^{\mathrm{BB2}}}{\alpha_{k+1}^{\mathrm{BB1}}} < \tau \\ \alpha_{k+1}^{\mathrm{BB1}}, \quad \tau = 1.1\,\tau, & \text{otherwise} \end{cases} \tag{5}$$

where $m_\alpha = 2$ and the initial value of τ is 0.5. The step-length rule introduced in [9] is obtained as follows. Denoting by G the $N_v \times m$ matrix defined by means of the last m gradients as

$$G = \left[D_{k-m+1}^{1/2}\widetilde{g}^{(k-m+1)}, \ldots, D_k^{1/2}\widetilde{g}^{(k)}\right], \qquad \widetilde{g}_j^{(k)} = \begin{cases} 0 & \text{if } x_j^{(k)} = 0, \\ \left[\nabla f(x^{(k)})\right]_j & \text{if } x_j^{(k)} > 0, \end{cases}$$

the matrix $\tilde{T} = [R \quad r]\Gamma R^{-1}$ is computed where R is the $m \times m$ upper triangular matrix obtained by the Cholesky factorization of $G^T G$, r is the solution of the linear system $R^T r = G^T D_{k+1}^{1/2}\widetilde{g}^{(k+1)}$ and Γ is an $(m+1) \times m$ bidiagonal matrix with non-zero entries $\Gamma_{j,j} = (\eta_{k-m+j}\alpha_{k-m+j})^{-1}$ and $\Gamma_{j+1,j} = -(\eta_{k-m+j}\alpha_{k-m+j})^{-1}$, $j = 1, \ldots, m$. From the upper Hessenberg matrix $\tilde{T}$ an $m \times m$ symmetric tridiagonal matrix T is defined by replacing the strictly upper triangle of $\tilde{T}$ by the transpose of its strictly lower triangle; the eigenvalues of T, θ_k, $k = 1, \ldots, m$, called Ritz-like values in [9], are used to define the step-lengths for the next m iterations: $\alpha_{k+j} = \frac{1}{\theta_j}$, $j = 1, \ldots, m$. The value $m = 3$ is used in our implementation.

4. Numerical results

In this section we present some preliminary numerical results performed with Matlab R2016b. We consider as the exact digital object x a 3D Shepp Logan test phantom constituted by $61 \times 61 \times 61$ voxels, whose central layer is shown in the left panel of Figure 1. The projection operator is obtained with the functions of the TVReg Matlab toolbox (from `http://www.imm.dtu.dk/~pcha/TVReg`) simulating a system where a source moves on a unit semi-sphere emitting X-ray cone beams from N_θ angles. In our experiments, N_θ has the values $19, 37, 55$, while the plain detector has 61×61 pixels, hence the matrix A in (1) has a rectangular shape, with less rows than columns. The projections obtained as $b = Ax$ are corrupted by random noise from Poisson distribution with level 10^9 and $\mathrm{bg} = 10^{-5}$. The regularization parameter λ is heuristically fixed as 0.03, while the smoothing parameter is $\beta = 0.01$. We solve problem (2) for different N_θ values and with four distinct implementations of the SGP algorithm, in order to test the effectiveness of the proposed strategies. The relative error ($RelErr$) between the exact object and the reconstruction, and the achieved values of the objective function are shown to evaluate the performance of the methods.

The plots in Figure 1 compare the errors (the middle panel) and the values of the objective function (the right panel) versus the iterations, for the different SGP implementations. The red and blue lines refer to the scaled methods with the step-length based on the BB rules (SGP_BB) and the Ritz-like values (SGP_R), respectively; the black and green lines denote the non-scaled methods (GP_BB and GP_R, respectively), that is, the SGP versions with D_k equal to the identity matrix at each iteration. Independently of the step-length rule, the scaling strategy accelerates the GP methods considerably, especially in the first iterations. In Figure 2 we show the reconstruction of the central layer obtained with the scaled methods, both after 20 and 1000 iterations: the quality of the SGP_R reconstruction after 20 iterations is indeed noticeable. In order to better compare the SGP_BB and SGP_R results after 20 iterations, we analyse the profiles highlighted in the image in the left panel of Figure 1: on the left of Figure 3 we show the plot related to the magenta line in layer 31; on the right we show the plot of the red single pixel over the 61 layers. We observe that the noise is suppressed almost everywhere and the small objects represented by narrow peaks are detected, even if the intensity of the signal is somewhere lower than the exact one. Furthermore, it is evident that the SGP_R gets a better reconstruction in few iterations and this is crucial for medical softwares that must provide reliable images in at most one minute. To conclude, in Table 2 we report some results obtained by the SGP_BB and SGP_R algorithms at 20 iterations, with a different number of source angles (consequently a different number of data); for $N_\theta = 19, 37, 55$, we report both the value of the objective function and the relative error. The table confirms that the choice of the step-length based on the Ritz-like values improves the algorithm speed in all the considered underdetermined cases, providing useful reduction of the reconstruction error.

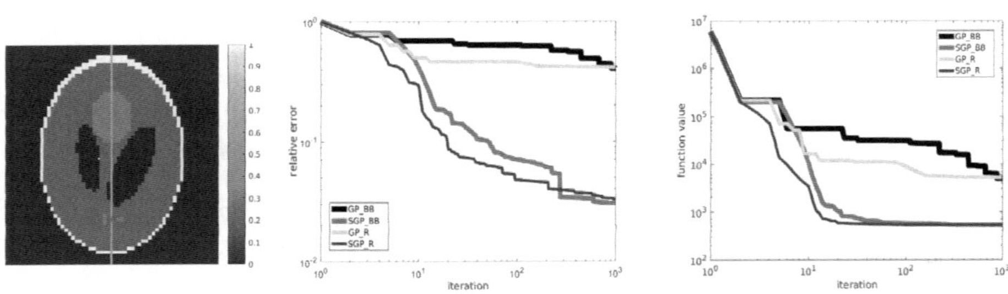

Figure 1: The central layer of the exact Shepp Logan phantom with the objects of interest (left), relative error vs iterations (middle) and objective function vs iterations (right), for $N_\theta = 37$.

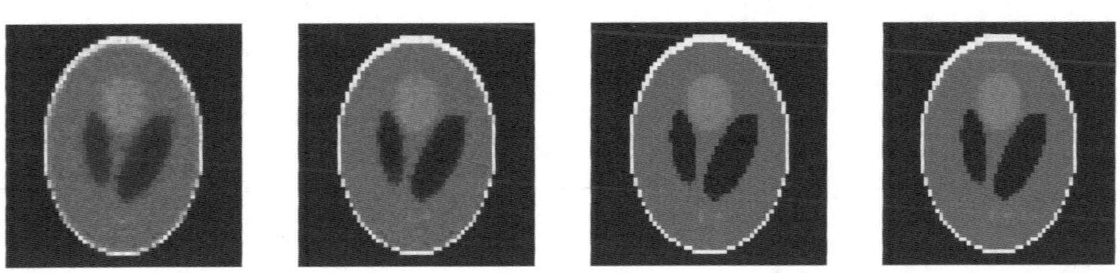

Figure 2: The central layer of the scaled reconstructions. From left to right: SGP_BB and SGP_R outputs in 20 iterations, SGP_BB and SGP_R outputs after 1000 iterations.

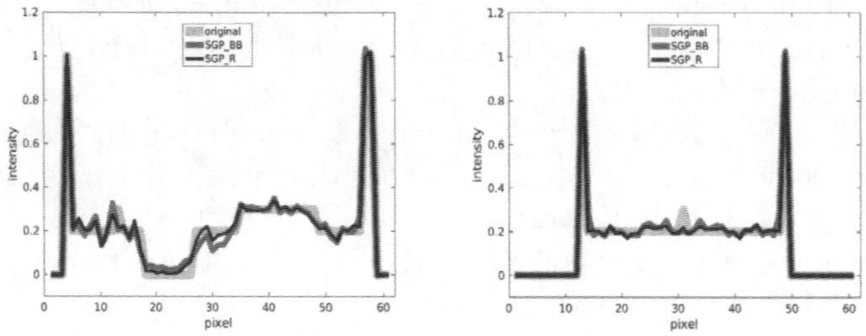

Figure 3: Vertical profile (on the left) and depth profile (on the right) of the central layer.

Table 2: Behavior of SGP_BB and SGP_R for different number of N_θ, at 20 iterations.

		$N_\theta = 19$		$N_\theta = 37$		$N_\theta = 55$	
		$f(x^{(k)})$	$RelErr$	$f(x^{(k)})$	$RelErr$	$f(x^{(k)})$	$RelErr$
k = 20	SGP_BB	765.109	0.2140	1055.09	0.1705	1310.77	0.1609
	SGP_R	553.704	0.1522	590.133	0.0856	661.329	0.0894

5. Conclusions

In this paper we have considered accelerated gradient projection methods based on both the use of scaling strategies and adaptive step-length selections for the reconstruction of 3D CT images from limited data. In particular, a generalization of the rule proposed in [10], defining the step-lengths by means of a small number of back gradients, improves the convergence in the first iterations with respect to a more traditional selection based on the alternation of the Barzilai-Borwein rules. This characteristic is essential in the real applications, where, despite the huge size of the problem, a good restored image must be computed in about one minute. We have presented preliminary results on a phantom, but the algorithm will also be tested on real images, in particular breast and head images that have features well detected by TV-like regularization functions.

Acknowledgments

This work has been partially supported by the Italian Institute GNCS - INdAM.

References

[1] Sidky E Y, Kao C M and Pan X 2006 *Journal of X-Ray Science and Technology* **14** 119–139
[2] Graff C and Sidky E 2015 *Applied Optics* **54** C23–C44
[3] Sidky E Y, Jørgensen J H and Pan X 2012 *Physics in Medicine and Biology* **57** 3065–3091
[4] Jensen T L, Jørgensen J H, Hansen P C and Jensen S H 2012 *BIT Numer Math* **52** 329–356
[5] Yu H and Wang G 2010 *Physics in Medicine and Biology* **55** 3905–3916
[6] Piccolomini E L and Morotti E 2016 *J. Algor. and Computat. Techn.* **10** 277–289
[7] Bonettini S, Zanella R and Zanni L 2009 *Inv. Probl.* **25**
[8] Frassoldati G, Zanni L and Zanghirati G 2008 *J. Ind. Manag. Optim.* **4** 299–312
[9] Porta F, Prato M and Zanni L 2015 *J. Sci. Comp.* **65** 895–919
[10] Fletcher R 2012 *Math. Program., Ser. A* **135** 413–436
[11] Epstein C L 2007 *Introduction to the Mathematics of Medical Imaging* 2nd ed (SIAM, Philadelphia)
[12] Oliveri G, Anselmi N and Massa A 2014 *IEEE Tr. Ant. Prop.* **62** 5157–5170
[13] Lantéri H, Roche M and Aime C 2002 *Inverse Probl.* **18** 1397–1419
[14] Zanella R, Boccacci P, Zanni L and Bertero M 2009 *Inv. Probl.* **25** 045010
[15] Barzilai J and Borwein J M 1988 *IMA J. Numer. Anal.* **8** 141–148
[16] Zhou B, Gao L and Dai Y H 2006 *Comput. Optim. Appl.* **35** 69–86
[17] Bonettini S and Prato M 2015 *Inverse Probl.* **31** 095008

Alternating Direction Implicit (ADI) schemes for a PDE-based image osmosis model

L. Calatroni[1], **C. Estatico**[2], **N. Garibaldi**[2], **S. Parisotto**[3]

[1] CMAP, Ecole Polytechnique 91128 Palaiseau Cedex, France
[2] Dipartimento di Matematica, Università di Genova, Via Dodecaneso 35, 16146, Genova, Italy
[3] Cambridge Centre for Analysis, Wilberforce Road, CB3 0WA, University of Cambridge, UK

E-mail: luca.calatroni@polytechnique.edu, estatico@dima.unige.it, garibaldi91nicola@gmail.com, sp751@cam.ac.uk

Abstract. We consider *Alternating Direction Implicit* (ADI) splitting schemes to compute efficiently the numerical solution of the PDE osmosis model considered by Weickert et al. in [10] for several imaging applications. The discretised scheme is shown to preserve analogous properties to the continuous model. The dimensional splitting strategy traduces numerically into the solution of simple tridiagonal systems for which standard matrix factorisation techniques can be used to improve upon the performance of classical implicit methods, even for large time steps. Applications to the shadow removal problem are presented.

1. Introduction

Several imaging tasks can be formulated as the problem of finding a reconstructed version of a degraded and possibly under-sampled image. Removing signal oscillations (*image denoising*) and optical aberrations (*image deblurring*) due to acquisition and transmission faults are common problems in applications such as medical imaging, astronomy, microscopy and many more. Further, the image interpolation problem of filling in missing or occluded parts of the image using the information from the surrounding areas is called *image inpainting* [3, 8]. More sophisticated imaging tasks such as shadow removal aim to decompose the acquired image into its component structures, such as cartoon and texture, or into the product of an "intrinsic" component not subject to light-changes (e.g. illumination) and its counterpart describing illumination changes only, see [4, 11]. For a given image f, the problem is

$$\text{find } u \quad \text{s.t.} \quad f(x,y) = l(x,y)u(x,y), \qquad (x,y) \in \Omega, \tag{1}$$

where $\Omega \subset \mathbb{R}^2$ denotes the image domain, $l(x,y)$ the illumination image and u the *reflectance* image, i.e. the intrinsic image not subject to illumination changes [11]. Due to the ill-posedness of the problem (1), in [11] a maximum-likelihood approach is used. In [9, 10], a PDE-based approach computing the solution of (1) as the stationary state of a linear parabolic PDE is considered. Due to its similarities with the physical process, such model is called *image osmosis.*

Image osmosis. Given a rectangular image domain $\Omega \subset \mathbb{R}^2$ with Lipschitz boundary $\partial\Omega$ and a positive initial gray-scale image $f : \Omega \to \mathbb{R}^+$, for a given vector field $\boldsymbol{d} : \Omega \to \mathbb{R}^2$ the linear image osmosis model proposed in [9, 10] is a drift-diffusion PDE which computes for every $t \in (0,T]$, $T > 0$ a regularised family of images $\{u(x,t)\}_{t>0}$ of f by solving:

$$\begin{cases} \partial_t u = \Delta u - \operatorname{div}(\boldsymbol{d}u) & \text{on } \Omega \times (0,T] \\ u(x,0) = f(x) & \text{on } \Omega \\ \langle \nabla u - \boldsymbol{d}u, \boldsymbol{n} \rangle = 0 & \text{on } \partial\Omega \times (0,T], \end{cases} \tag{2}$$

where $\langle \cdot, \cdot \rangle$ denotes the Euclidean scalar product and $\boldsymbol{n}$ the outer normal vector on $\partial\Omega$. The extension to RGB images is straightforward by letting the process evolve for each colour channel. In [10, Proposition 1] the authors prove that any solution of (2) preserves the average gray value and the non-negativity at any time $t > 0$. Furthermore, by setting $\boldsymbol{d} := \boldsymbol{\nabla}(\ln v)$ for a reference image $v > 0$, the steady state w of (2) turns out to be a minimiser of a quadratic energy functional and can be expressed as a rescaled version of v via the formula $w(x) = \frac{\mu_f}{\mu_v} v(x)$, where μ_f and μ_v are the average of f and v over Ω, respectively.

In [10] the osmosis model (2) is used as a non-symmetric variant of diffusion for several visual computing applications (data compression, face cloning and shadow removal). A general fully discrete framework for this problem is studied in [9] where, under some standard conditions on the semi-discretised finite differences operators used, the conservation properties described above are shown to be still valid upon discretisation. We synthesise these results as follows:

Theorem 1 ([9]). *For a given $\boldsymbol{f} \in \mathbb{R}^N_+$, consider the semi-discretised linear osmosis problem:*

$$\boldsymbol{u}(0) = \boldsymbol{f}, \qquad \boldsymbol{u}'(t) = \boldsymbol{A}\boldsymbol{u}(t), \qquad t > 0, \tag{3}$$

where $\boldsymbol{A} \in \mathbb{R}^{N \times N}$ is an irreducible (non-symmetric) matrix with zero-column sum and non-negative off-diagonal entries. For $k \geq 0$, consider the time-discretisation schemes $\boldsymbol{u}^{k+1} = \boldsymbol{P}\boldsymbol{u}^k$:

$$\boldsymbol{P} := (\boldsymbol{I} + \tau\boldsymbol{A}), \qquad \textit{for } \tau < (\max|a_{i,i}|)^{-1}; \qquad \textbf{\textit{Forward Euler}} \ \textbf{(F.E.)}$$

$$\boldsymbol{P} := (\boldsymbol{I} - \tau\boldsymbol{A})^{-1}, \qquad \textit{for any } \tau > 0. \qquad \textbf{\textit{Backward Euler}} \ \textbf{(B.E.)}$$

Then, in both cases, $\boldsymbol{P}$ is an irreducible, non-negative matrix with strictly positive diagonal entries and unitary column sum and for $\boldsymbol{u}(t)$ the following properties hold true:

(i) For every $t > 0$, the evolution preserves positivity and the average gray value of $\boldsymbol{f}$;

(ii) The unique steady state is the eigenvector of $\boldsymbol{P}$ associated to eigenvalue 1.

For large $\tau > 0$, the numerical realisation via B.E. requires the inversion of a non-symmetric penta-diagonal matrix, which may be highly costly for large images. In this paper we solve problem (3) using accurate dimensional semi-implicit splitting methods, [5]. Similarly, fully implicit splitting schemes are considered in [6] and applied to large images for cultural heritage.

ADI splitting methods. Given a domain $\Omega \in \mathbb{R}^s$, a *dimensional splitting* method decomposes the semi-discretised operator $\boldsymbol{A}$ of initial boundary value problem (3) into the sum:

$$\boldsymbol{A} = \boldsymbol{A_0} + \boldsymbol{A_1} + \ldots + \boldsymbol{A_s} \tag{4}$$

such that the components $\boldsymbol{A_j}$, $j = 1, \ldots, s$, encode the linear action of $\boldsymbol{A}$ along the space direction $j = 1, \ldots, s$, respectively. The term $\boldsymbol{A_0}$ may contain additional contributes coming from mixed directions and non-stiff nonlinear terms. *Alternating directional implicit* (ADI) schemes are time-stepping methods that treat the unidirectional components $\boldsymbol{A_j}, j \geq 1$ implicitly and the $\boldsymbol{A_0}$ component, if present, explicitly in time. Having in mind a standard finite difference space discretisation, this splitting idea translates into reducing the s-dimensional original problem to s one-dimensional problems.

For imaging applications $s = 2$. Let $u_{i,j}$ denote an approximation of u in the grid of size h at point $((i - \frac{1}{2})h, (j - \frac{1}{2})h)$; the discretisation of (2) considered in [9, 10] reads:

$$\begin{aligned} u'_{i,j} &= \frac{u_{i+1,j} - 2u_{i,j} + u_{i-1,j}}{h^2} - \frac{1}{h}\left(d_{1,i+\frac{1}{2},j}\frac{u_{i+1,j} + u_{i,j}}{2} - d_{1,i-\frac{1}{2},j}\frac{u_{i,j} + u_{i-1,j}}{2}\right) \\ &+ \frac{u_{i,j+1} - 2u_{i,j} + u_{i,j-1}}{h^2} - \frac{1}{h}\left(d_{2,i,j+\frac{1}{2}}\frac{u_{i,j+1} + u_{i,j}}{2} - d_{2,i,j-\frac{1}{2}}\frac{u_{i,j} + u_{i,j-1}}{2}\right) =: (\boldsymbol{A}_1 + \boldsymbol{A}_2)\boldsymbol{u}(t) \end{aligned} \tag{5}$$

which can be splitted into the sum of $\boldsymbol{A_1}$ and $\boldsymbol{A_2}$ where no mixed and/or nonlinear terms appear ($\boldsymbol{A_0} = \boldsymbol{0}$). Following [5], we recall now the two ADI schemes considered in this paper.

The Peaceman-Rachford scheme. The first method considered is the second-order accurate *Peaceman-Rachford* ADI scheme. For every $k \geq 0$ and time-step $\tau > 0$, compute an approximation $\boldsymbol{u}^{k+1}$ via the updating rule:

$$\begin{cases} \boldsymbol{u}^{k+1/2} = \boldsymbol{u}^k + \frac{\tau}{2}\boldsymbol{A_1}\boldsymbol{u}^k + \frac{\tau}{2}\boldsymbol{A_2}\boldsymbol{u}^{k+1/2}, \\ \boldsymbol{u}^{k+1} = \boldsymbol{u}^{k+1/2} + \frac{\tau}{2}\boldsymbol{A_1}\boldsymbol{u}^{k+1} + \frac{\tau}{2}\boldsymbol{A_2}\boldsymbol{u}^{k+1/2}, \end{cases} \tag{6}$$

where F.E. and B.E. are applied alternatively and in a symmetric way.

The Douglas scheme. A more general ADI decomposition accommodating also the general case $\boldsymbol{A_0} \neq \boldsymbol{0}$ in (4) is the *Douglas* method. For $k \geq 0$, $\tau > 0$ and $\theta \in [0, 1]$, the updating rule reads

$$\begin{cases} \boldsymbol{y}^0 = \boldsymbol{u}^k + \tau \boldsymbol{A}\boldsymbol{u}^k \\ \boldsymbol{y}^j = \boldsymbol{y}^{j-1} + \theta\tau(\boldsymbol{A_j}\boldsymbol{y}^j - \boldsymbol{A_j}\boldsymbol{u}^k), \qquad j = 1, 2 \\ \boldsymbol{u}^{k+1} = \boldsymbol{y}^2. \end{cases} \tag{7}$$

In words, the numerical approximation in each time step is computed by applying at first a F.E. predictor and then it is stabilised by intermediate steps where just the unidirectional components $\boldsymbol{A_j}$ of the splitting (4) appear, weighted by θ whose size balances the implicit/explicit behaviour of these steps. The time-consistency order of the scheme is equal to two for $\boldsymbol{A_0} = \boldsymbol{0}$ and $\theta = 1/2$, and it is of order one otherwise.

For $s = 2$ both schemes (6) and (7) are unconditionally stable, see [5]. However, for large time steps, the time-accuracy may suffer due to the presence of the explicit steps.

Scope of the paper. In this work, we apply schemes (6) and (7) to solve (3) and show that similar conservation properties as in Theorem 1 hold. Compared to standard implicit methods used in [9, 10], such schemes renders a much more efficient and accurate numerical solution.

2. ADI methods for image osmosis

We show that ADI schemes (6) and (7) show similar conservation properties as in Theorem 1.

Proposition 1. *Let* $\boldsymbol{f} \in \mathbb{R}^N_+$ *and* $\tau < 2\Big(\max\Big\{\max|a^1_{i,i}|, \max|a^2_{i,i}|\Big\}\Big)^{-1}$. *Then, the Peaceman-Rachford scheme (6) on the splitting (5) preserves the average gray value, the positivity and converges to a unique steady state.*

Proof. We write the Peaceman-Rachford (6) iteration as

$$\boldsymbol{u}^{k+1} = \left(\boldsymbol{I} - \frac{\tau}{2}\boldsymbol{A_1}\right)^{-1}\left(\boldsymbol{I} + \frac{\tau}{2}\boldsymbol{A_2}\right)\left(\boldsymbol{I} - \frac{\tau}{2}\boldsymbol{A_2}\right)^{-1}\left(\boldsymbol{I} + \frac{\tau}{2}\boldsymbol{A_1}\right)\boldsymbol{u}^k,$$

and observe that for $i = 1, 2$ the matrices $\boldsymbol{P_i^-} := \left(\boldsymbol{I} - \frac{\tau}{2}\boldsymbol{A_i}\right)^{-1}$ and $\boldsymbol{P_i^+} := \left(\boldsymbol{I} + \frac{\tau}{2}\boldsymbol{A_i}\right)$, are non-negative and irreducible with unitary column sum and positive diagonal entries being each $\boldsymbol{A_i}$ a one-dimensional implicit/explicit discretised osmosis operator satisfying the assumptions of Theorem 1. Therefore, at every implicit/explicit half-step and using the restriction on τ, the average gray-value and positivity are conserved. Furthermore, the unique steady state is the eigenvector of the operator $\boldsymbol{P} := \boldsymbol{P_1^-}\boldsymbol{P_2^+}\boldsymbol{P_2^-}\boldsymbol{P_1^+}$ associated to the eigenvalue to one. □

The following lemma is useful to prove a similar result for the Douglas scheme (7).

Lemma 1. *If $\boldsymbol{C} = (c_{i,j}) \in \mathbb{R}^{N\times N}$ and $\boldsymbol{D} = (d_{i,j}) \in \mathbb{R}^{N\times N}$ s.t. $\sum_{i=1}^N c_{i,j} =: c$ and $\sum_{i=1}^N d_{i,j} =: d$ for every $j = 1, \ldots, N$, then, the matrix $\boldsymbol{B} := \boldsymbol{CD}$ has column sum equal to cd.*

Proof. Writing each element $b_{i,j}$ in terms of the elements of $\boldsymbol{C}$ and $\boldsymbol{D}$, we have that for every j:

$$\sum_{i=1}^N b_{i,j} = \sum_{i=1}^N\sum_{k=1}^N c_{i,k}d_{k,j} = \sum_{k=1}^N\left(d_{k,j}\sum_{i=1}^N c_{i,k}\right) = \sum_{k=1}^N\left(d_{k,j}\cdot c\right) = c\sum_{k=1}^N d_{k,j} = cd. \qquad \square$$

Proposition 2. *Let $f \in \mathbb{R}_+^N$, $\tau > 0$ and $\theta \in [0, 1]$. Then, the Douglas scheme (7) applied to the splitted semi-discretised scheme (5) preserves the average gray value.*

Proof. For every $k \geq 0$, we write the Douglas iteration (7) as

$$\begin{aligned}\boldsymbol{u}^{k+1} &= (\boldsymbol{I} - \theta\tau\boldsymbol{A_2})^{-1}\left[(\boldsymbol{I} - \theta\tau\boldsymbol{A_1})^{-1}\left((\boldsymbol{I} - \theta\tau\boldsymbol{A_1}) + \tau\boldsymbol{A}\right) - \theta\tau\boldsymbol{A_2}\right]\boldsymbol{u}^k \\ &= \left(\boldsymbol{I} + \tau(\boldsymbol{I} - \theta\tau\boldsymbol{A_2})^{-1}(\boldsymbol{I} - \theta\tau\boldsymbol{A_1})^{-1}\boldsymbol{A}\right)\boldsymbol{u}^k =: \left(\boldsymbol{I} + \tau\boldsymbol{P_2}\boldsymbol{P_1}\boldsymbol{A}\right)\boldsymbol{u}^k\end{aligned}$$

where both $\boldsymbol{P_1}$ and $\boldsymbol{P_2}$ are non-negative, irreducible, with unitary column sum and strictly positive diagonal entries by Theorem 1, while $\boldsymbol{A} = \boldsymbol{A_1} + \boldsymbol{A_2}$ is zero-column sum being the standard discretised osmosis operator (5). By Lemma 1, the operator $\boldsymbol{B} := \boldsymbol{P_2}\boldsymbol{P_1}\boldsymbol{A}$ has zero column sum and the operator $\boldsymbol{P} := \boldsymbol{I} + \tau\boldsymbol{B}$ has unitary column sum. Thus, for every $k \geq 0$:

$$\frac{1}{N}\sum_{i=1}^N u_i^{k+1} = \frac{1}{N}\sum_{i=1}^N\sum_{j=1}^N p_{i,j}u_j^k = \frac{1}{N}\sum_{j=1}^N\left(\sum_{i=1}^N p_{i,j}\right)u_j^k = \frac{1}{N}\sum_{j=1}^N u_j^k. \qquad \square$$

Remark 1. *There is no guarantee that the off-diagonal entries of $\boldsymbol{B}$ are non-negative. Therefore, it is not possible to apply directly Theorem 1 to conclude that the iterates $\{\boldsymbol{u}^k\}_{k\geq 0}$ remain positive and converge to a unique steady state. However, our numerical tests suggest that both properties remain valid. A rigorous proof of these properties is left for future research.*

3. Numerical results

We present in Algorithm 1 and 2 the pseudocode for the Peaceman-Rachford (6) and the Douglas (7) schemes, respectively, when applied to images of size $m \times n$ pixels, with $N := mn$. We recall that in [9, 10], the non-split problem is solved iteratively by B.E. using `BiCGStab`, whose performance is influenced by the accuracy and maximum iterations required.

Due to the structure of the ADI matrices $\boldsymbol{A_1}$ and $\boldsymbol{A_2}$, a tridiagonal `LU` factorisation can be used for improved efficiency after a permutation on matrix $\boldsymbol{A_2}$ based on [7] to reduce the bandwidth to one: this is convenient for images with large m, where standard methods become prohibitively expensive. Without splitting, one LU factorisation plus $T\tau^{-1}$ system resolutions for the penta-diagonal matrix $\boldsymbol{A}$, with lower and upper bandwidth equal to m, costs $\mathcal{O}(2m^3n + 4T\tau^{-1}m^2n)$ *flops* overall. Using splitting, the computational cost is reduced to $\mathcal{O}(6(mn-1) + T\tau^{-1}(10mn-8))$ *flops* because of the use of tridiagonal 1-bandwidth matrices. For the same reason, the inverse operators $\boldsymbol{A_1}$ and $\boldsymbol{A_2}$ require less storage space than the full operator $\boldsymbol{A}$ while the inverse of $\boldsymbol{A}$ has significantly more non-zeros entries. Note that since τ is constant, all the LU factorisations can be computed only once before the main loop.

Algorithm 1: LU Peaceman-Rachford

Input : u^0, $\mathtt{A} = \mathtt{A}_1 + \mathtt{A}_2$;
Parameters : τ, T;
foreach c *in* color channel **do**
  $\mathtt{p}_c = \mathtt{symrcm}(\mathtt{A}_{2c})$;
  $\mathtt{E1}_c = (\mathtt{I} + 0.5\,\tau\,\mathtt{A}_{1c})$;
  $\mathtt{E2}_c = (\mathtt{I} + 0.5\,\tau\,\mathtt{A}_{2c}(\mathtt{p}_c\,;\,\mathtt{p}_c))$;
  $[\mathtt{L1}_c, \mathtt{U1}_c] = \mathtt{lu}(\mathtt{I} - 0.5\,\tau\,\mathtt{A}_{1c})$;
  $[\mathtt{L2}_c, \mathtt{U2}_c] = \mathtt{lu}(\mathtt{I} - 0.5\,\tau\,\mathtt{A}_{2c}(\mathtt{p}_c\,;\,\mathtt{p}_c))$.
  for $k = 0$ **to** $T - \tau$ **do**
    $z_1 = \mathtt{E1}_c\,u_c^k$;
    $y_1(\mathtt{p}_c) = \mathtt{U2}_c \setminus (\mathtt{L2}_c \setminus z_1(\mathtt{p}_c))$;
    $z_2(\mathtt{p}_c) = \mathtt{E2}_c\,y_1(\mathtt{p}_c)$;
    $u_c^{k+\tau} = \mathtt{U1}_c \setminus (\mathtt{L1}_c \setminus z_2)$;
return u^T.

Algorithm 2: LU-Douglas

Input : u^0, $\mathtt{A} = \mathtt{A}_1 + \mathtt{A}_2$;
Parameters : τ, T, $\theta \in [0,1]$;
foreach c *in* color channel **do**
  $\mathtt{p}_c = \mathtt{symrcm}(\mathtt{A}_{2c})$;
  $\mathtt{E1}_c = \tau\,\mathtt{A}_{1c}$; $\mathtt{E2}_c = \tau\,\mathtt{A}_{2c}(\mathtt{p}_c\,;\,\mathtt{p}_c)$
  $\mathtt{T1}_c = \tau\,\theta\,\mathtt{A}_{1c}$; $\mathtt{T2}_c = \tau\,\theta\,\mathtt{A}_{2c}(\mathtt{p}_c\,;\,\mathtt{p}_c)$;
  $[\mathtt{L1}_c, \mathtt{U1}_c] = \mathtt{lu}(\mathtt{I} - \tau\,\theta\,\mathtt{A}_{1c})$;
  $[\mathtt{L2}_c, \mathtt{U2}_c] = \mathtt{lu}(\mathtt{I} - \tau\,\theta\,\mathtt{A}_{2c}(\mathtt{p}_c\,;\,\mathtt{p}_c))$.
  for $k = 0$ **to** $T - \tau$ **do**
    $y_{01} = \mathtt{E1}_c\,u_c^k$; $y_{02}(\mathtt{p}_c) = \mathtt{E2}_c\,u_c^k(\mathtt{p}_c)$;
    $z_1 = u_c^k + y_{01} + y_{02} - \mathtt{T1}_c\,u_c^k$;
    $y_1 = \mathtt{U1}_c \setminus (\mathtt{L1}_c \setminus z_1)$;
    $z_2(\mathtt{p}_c) = y_1(\mathtt{p}_c) - \mathtt{T2}_c\,u_c^k$;
    $u_c^{k+\tau}(\mathtt{p}_c) = \mathtt{U2}_c \setminus (\mathtt{L2}_c \setminus z_2(\mathtt{p}_c))$;
return u^T.

As a toy example, we let the osmosis model (2) evolve towards a known steady state starting from an initial constant image. In Table 1 we compare the numerical solutions at $T = 5000$ for different fixed time-steps τ and values of θ. The benchmark solution $\bar{u}$ is computed using Exponential Integrators [2], which for linear equations are exact-in-time solvers. Similarly as in [9], the approach is solved without splitting by `BiCGStab` (with parameters `tol=1e-07` and `maxiter=3e+05`). For comparison, we test the direct LU factorisation applied to the full problem. The order of the methods for the relative Root Mean Square (RMS) error (rRMSE) $\mathtt{rms}(u_T - \bar{u})/\mathtt{rms}(\bar{u})$ are plotted in Figure 1a. For large τ, ADI methods are up to 10× faster than methods solving the problem without any splitting.

Table 1: Full vs. ADI splitting p-order methods at T=5000. Input: 240 × 250 pixel image.

			$\tau = 0.1$		$\tau = 1$		$\tau = 10$		$\tau = 100$	
Method	θ	p	Time	rRMSE	Time	rRMSE	Time	rRMSE	Time	rRMSE
`BiCGStab` (5 d.)	1	1	584.90 s.	3.18e-06	104.86 s.	7.74e-06	25.73 s.	6.82e-05	13.91 s.	6.64e-04
LU (5 d.)	1	1	**3165.61 s.**	3.60e-07	310.01 s.	3.60e-06	33.83 s.	3.60e-05	3.07 s.	3.60e-04
`Douglas` ADI	1	1	529.59 s	3.60e-07	57.36 s.	3.61e-06	6.12 s.	3.67e-05	0.62 s.	1.47e-02
`BiCGStab` (5 d.)	0.5	2	600.70 s.	1.19e-06	97.54 s.	2.94e-05	33.72 s.	4.45e-06	20.60 s.	2.05e-02
LU (5 d.)	0.5	2	**3652.58 s.**	7.93e-12	321.95 s.	3.11e-10	32.47 s.	3.11e-08	3.20 s.	2.05e-02
`Douglas` ADI	0.5	2	523.85 s.	3.56e-11	53.06 s.	3.56e-09	5.18 s.	3.56e-07	0.52 s.	2.32e-02
`P.-R.` ADI	-	2	463.68 s.	6.32e-11	36.35 s.	6.32e-09	3.56 s.	6.33e-07	0.38 s.	8.10e-02

In Figure 1b the ADI numerical solution of the shadow-removal problem [10] is shown, starting from a shadowed image f and setting $\boldsymbol{d} = \boldsymbol{\nabla}(\ln f) = 0$ on the shadow boundary. In order to

IOP Conf. Series: Journal of Physics: Conf. Series **904** (2017) 012014 doi:10.1088/1742-6596/904/1/012014

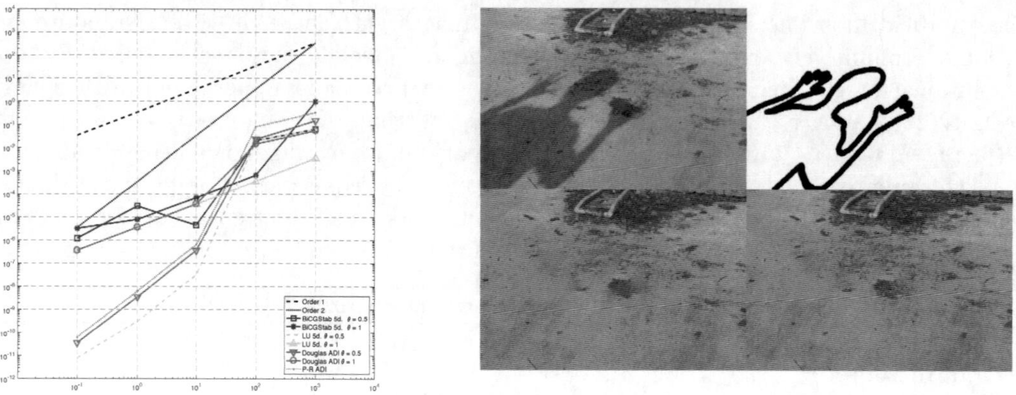

(a) Loglog order plot: τ vs. rRMSE (T=5000).

(b) Shadow removal and inpainting.

Figure 1: 1a: Relative RMS error at $T = 5000$ for several τ. 1b: Shadow removal inputs (top); Douglas ADI solution and result after non-local inpainting [1] (bottom). Parameters: $\tau = 10$; $\theta = 0.5$. Image from: `http://aqua.cs.uiuc.edu/site/projects/shadow.html`.

correct the diffusion artefacts, an inpainting correction (e.g. [1]) is applied.

4. Conclusions

In this paper, we consider the *image osmosis* model proposed in [9, 10] for imaging applications. We proposed dimensional splitting approaches to decompose the discretised 2D operator into two tridiagonal operators. We proved that the splitting numerical schemes preserve analogous properties holding in the continuous model. Numerically, these methods are efficient and accurate. We tested the proposed methods for a shadow removal problem, with an inpainting post-processing step. Future research directions include the rigorous convergence proof of the discrete models to the continuum one as $N \to \infty$ as well as a rigorous stability and convergence analysis for the Douglas method (7).

Acknowledgments

LC acknowledges the joint ANR/FWF Project “Efficient Algorithms for Nonsmooth Optimization in Imaging” (EANOI) FWF n. I1148 / ANR-12-IS01-0003. CE acknowledges the support of GNCS-INDAM and PRIN 2012 N. 2012MTE38N. SP acknowledges UK EPSRC grant EP/L016516/1.

References

[1] Arias P, Facciolo G, Caselles V and Sapiro G 2011 *ICJV* **93** 319-347
[2] Al-Mohy and A.H. Higham N J 2011 *SIAM J. Sci. Comput.* **33** 488-511
[3] Aubert G and Kornprobst P 2006 *Mathematical problems in image processing: Partial Differential Equations and the Calculus of Variations* 2nd ed Applied Mathematical Sciences (Springer) ISBN 9780387322001
[4] Finlayson G, Hordley S, Lu C and Drew M 2002 *ECCV* in LNCS (Springer) **2353** 823-836
[5] Hundsdorfer W and Verwer J 2003 *Numerical solution of time-dependent advection-diffusion-reaction equations* Computational Mathematics (Springer) ISBN 9783642057076
[6] Parisotto S, Calatroni L and Daffara C 2017, arXiv preprint: 1704.04052.
[7] Saad Y.2003 *Iterative Methods for Sparse Linear Systems* 2nd ed (SIAM) ISBN 9780898715347
[8] Schönlieb C-B 2015 *Partial Differential Equation Methods for Image Inpainting* Cambridge Monographs on Applied and Computational Mathematics (CUP) ISBN 9781107001008
[9] Vogel O, Hagenburg K, Weickert and J Setzer S 2013 *SSVM* in LNCS (Springer) **7893** 368-379
[10] Weickert J, Hagenburg and K BreußVogel O 2013 *EMMCVPR* in LNCS (Springer) **8081** 26-39
[11] Weiss Y 2001 *ICCV* in IEEE Proceedings on ICCV **2** 68-75

A low-rank approach to off-the-grid sparse deconvolution

Paul Catala[1], **Vincent Duval**[2,3], **and Gabriel Peyré**[1,4]

[1] DMA, ENS, 45 rue d'Ulm, 75005 Paris
[2] Mokaplan, Inria Paris, 2 rue Simone Iff, 75012 Paris
[3] CEREMADE, Univ. Paris-Dauphine, Place du Maréchal de Lattre de Tassigny, 75016 Paris
[4] CNRS, UMR 8553

E-mail: pcatala@dma.ens.fr, vincent.duval@inria.fr, gabriel.peyre@ens.fr

Abstract. We propose a new solver for the sparse spikes deconvolution problem over the space of Radon measures. A common approach to off-the-grid deconvolution considers semidefinite (SDP) relaxations of the total variation (the total mass of the measure) minimization problem. The direct resolution of this SDP is however intractable for large scale settings, since the problem size grows as f_c^{2d} where f_c is the cutoff frequency of the filter. Our first contribution introduces a penalized formulation of this semidefinite lifting, which has low-rank solutions. Our second contribution is a conditional gradient optimization scheme with non-convex updates. This algorithm leverages both the low-rank and the convolutive structure of the problem, resulting in an $O(f_c^d \log f_c)$ complexity per iteration. Numerical simulations are promising and show that the algorithm converges in exactly k steps, k being the number of Diracs composing the solution.

1. Introduction

1.1. Sparse spikes deconvolution

Superresolution problems aim at accurately recovering a sparse signal from low-resolution and possibly noisy measurements. This is an important challenge in fields such as medical imaging, microscopy or astronomy, where it may be crucial to overcome the physical limits of the resolving power of sensing devices.

Formally, we want to retrieve a d-dimensional discrete Radon measure $\mu_0 = \sum_{k=1}^{r} a_i \delta_{x_i}$ ($a_i \in \mathbb{R}$, $x_i \in \mathbb{T}^d$, where $\mathbb{T} = \mathbb{R}/\mathbb{Z}$ is the torus) given the discrete observations

$$y = \Phi\mu_0 + \sigma w \in \mathbb{C}^n,$$

where w is some normalized noise, and Φ is a known filter. In this work we focus on the case where Φ is a low-pass filter of cutoff frequency f_c: for any measure $\mu \in \mathcal{M}(\mathbb{T}^d)$, we have

$$\Phi\mu = \int_{\mathbb{T}^d} \phi(x)\mathrm{d}\mu(x), \quad \text{where} \quad \phi(x) = \left[e^{2i\pi\langle k,\, x\rangle}\right]_{|k|\le f_c}. \tag{1}$$

1.2. Beurling LASSO

Although this inverse problem is severely ill-posed, sparse estimates can be found by solving the following minimization problem, known as the BLASSO [1]:

$$\mu_\lambda \in \arg\min_{\mu\in\mathcal{M}(\mathbb{T}^d)} \frac{1}{2\lambda}\|y - \Phi\mu\|^2 + |\mu|(\mathbb{T}), \tag{$\mathcal{P}_\lambda$}$$

where the total variation norm is defined as

$$|\mu|(\mathbb{T}) = \sup\left\{\int_{\mathbb{T}} \eta \mathrm{d}\mu \; ; \; \eta \in \mathcal{C}(\mathbb{T}), \quad \|\eta\|_\infty \leq 1\right\},$$

and naturally extends the ℓ^1-norm of finite-dimensional vectors to the continuous setting of measures. This grid-free approach offers beneficial mathematical insight on the problem, leading to sharp criteria for stable spikes recovery [2, 3, 4]. However, the infinite-dimensionality of $(\mathcal{P}_\lambda)$ poses a numerical challenge.

The dual formulation plays an important role in the practical resolution of the BLASSO. It aims at the recovery of a trigonometric polynomial $\Phi^* p_\lambda$ where p_λ is the solution of the program

$$p_\lambda = \max_{p \in \mathbb{C}^{n^d}} \left\{\frac{1}{2}\|\frac{y}{\lambda} - p\|^2 \; ; \; \|\Phi^* p\|_\infty \leq 1\right\}. \tag{$\mathcal{D}_\lambda$}$$

In the case where Φ is a low-pass filter, knowing this dual certificate then enables accurate recovery using root-finding techniques [3], since primal-dual optimality relations ensure that the support of μ_λ is contained in the set $\{x \; ; \; |\Phi^* p_\lambda(x)| = 1\}$, see Figure 2. In practice, we recover the support by extracting the double roots of $1 - |\Phi^* p_\lambda|^2$. Amplitudes are then deduced using the relation $\Phi\mu_\lambda = y - \lambda p_\lambda$ (see Algorithm 2).

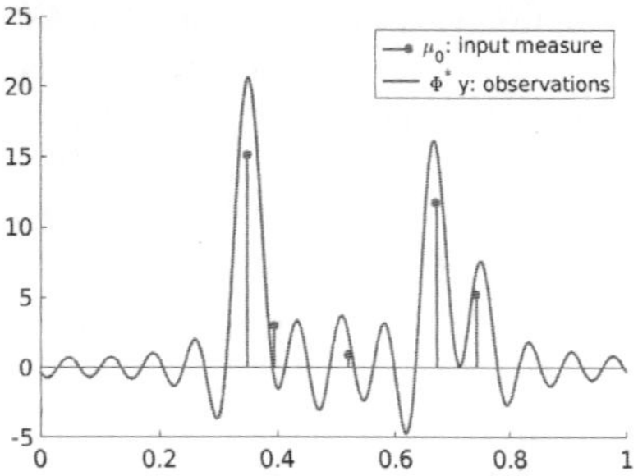

Figure 1: Coarse and noisy observations obtained from 5 spikes ($\sigma = .002$, $f_c = 13$).

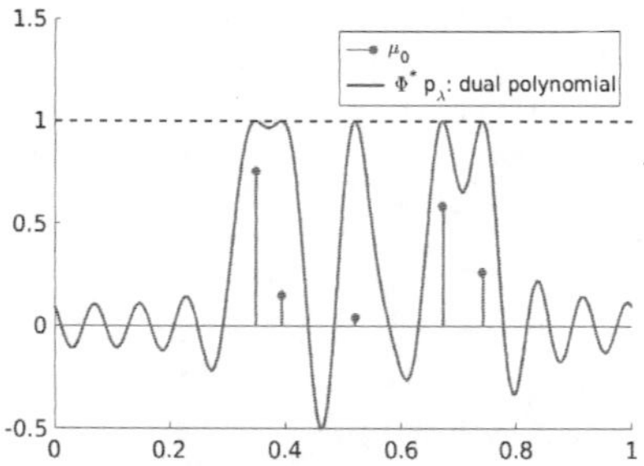

Figure 2: Dual polynomial $\eta_\lambda = \Phi^* p_\lambda$, with the same initial measure.

1.3. Previous works

Early works on superresolution considered a discrete approximation of the problem, restraining the search to measures located on a grid. This parameterization enabled the use of basis pursuit methods [5], also known as LASSO.

However, this approximation typically leads to imperfect solutions for thin grids [4], and better results as well as refined mathematical analysis are achieved using a grid-free setting. The support of the measure may then be iteratively estimated using conditional gradient methods [2, 6], or recovered through the dual polynomial, which can be computed in a robust manner using semidefinite relaxations [7, 3, 8].

Although we focus here on ℓ^1-regularization techniques, there is also a vast literature on non-convex superresolution schemes, such as Prony's method [9], MUSIC [10] or FRI-based methods [11], among others.

2. Low-rank SDP relaxation

Although $(\mathcal{P}_\lambda)$ and $(\mathcal{D}_\lambda)$ can be used for any degradation operator Φ, we assume in the following sections that the latter is of the form (1).

To solve these problems, one may consider their semidefinite relaxations. SDP hierarchies were introduced by Lasserre [12], and rely on the one hand on a re-parameterization of measures in terms of moment sequences, which remedies the infinite-dimensionality of $(\mathcal{P}_\lambda)$, and on sum-of-squares representations of nonnegative polynomials on the other, which allows to tackle the ℓ^∞-constraint of $(\mathcal{D}_\lambda)$. Although the solutions of these hierarchies have only been proven to converge weakly toward the solution of their original problems, the interest of this method comes from the fact that finite convergence is very frequent in practice.

The Lasserre hierarchy was originally intended for real measures and real polynomials. As our problem involves measures on the torus and trigonometric polynomials, we refer to [13] for the derivations of our semidefinite formulations.

2.1. *SDP relaxation*

The semidefinite relaxation of $(\mathcal{P}_\lambda)$ at order $(m-n+1)$ reads:

$$\mathcal{R}_\lambda \in \underset{u,z,\tau}{\operatorname{argmin}} \; h(\mathcal{R}) \overset{\text{def.}}{=} u_0 + \tau + \frac{1}{2}\|\frac{y}{\lambda} + 2z\|^2 \qquad (\widehat{\mathcal{P}}_\lambda^{(m)})$$

$$\text{s.t.} \quad \begin{cases} (a) \quad \mathcal{R} = \begin{bmatrix} R & z \\ z^* & \tau \end{bmatrix} \succeq 0, \\ (b) \quad R = \sum_{\|\boldsymbol{k}\|_\infty < m} u_{\boldsymbol{k}} \Theta_{\boldsymbol{k}}, \quad \Theta_{\boldsymbol{k}} \in \mathbb{C}^{m^d \times m^d} \end{cases}$$

where $n = 2f_c + 1$, $m \geq n$, $\boldsymbol{k} = (k_1, \ldots, k_d)$, and $\Theta_{\boldsymbol{k}} = \Theta_{k_d} \otimes \ldots \otimes \Theta_{k_1}$. Here Θ_{k_j} denotes the Toeplitz matrix with ones on its k_j-th diagonal and zeros everywhere else, and $\otimes$ stands for the Kronecker product. This formulation follows from the bounded real lemma [13]. The coefficients p_λ of the dual polynomial may then be retrieved from $\mathcal{R}_\lambda$ through the simple optimality relation

$$p_\lambda = \frac{y}{\lambda} + 2z_\lambda. \qquad (2)$$

In 1-D, the relaxation is exact with $m = n$, meaning that $(\widehat{\mathcal{P}}_\lambda^{(m)})$ and $(\mathcal{P}_\lambda)$ have the same solution (it results from Caratheodory-Toeplitz theorem [7], or from Fejér-Riesz theorem [13]). In d-D with $d > 1$, the relaxation is not tight in general, but numerical simulations tend to show that low relaxation orders are sufficient. In particular, as mentioned in [8, Section 4], for $d = 2$, the hierarchy is known to collapse, but with no a priori information on the order m at which it occurs.

As it appears, the size of the above semidefinite program is m^{2d}, with $m \geq 2f_c+1$. Therefore, interior points methods are limited when $d > 1$, or even in 1-D for large values of f_c.

2.2. *Low-rank structure*

An inspection of the proof in [7, Proposition 2.1] reveals that $(\widehat{\mathcal{P}}_\lambda^{(m)})$ in the univariate case $(d = 1)$ admits a solution whose rank is bounded by the number of spikes in the measure solving $(\mathcal{P}_\lambda)$. Furthermore, this result seems to hold in the d-dimensional case [14].

The search space of $(\widehat{\mathcal{P}}_\lambda^{(m)})$ may thus be restricted to matrices of rank at most r, greatly reducing the spatial complexity. However, the intersection between the manifold of fixed rank matrices and the affine space defined by (b) quickly becomes unanalyzable when $r > 1$ or $n > 2$, and non-convex optimization schemes on this search space seem difficult to implement. Instead, we propose to smooth the geometry of the problem by penalizing the affine constraint (b).

3. Toeplitz penalization

To overcome the difficulty induced by constraint (b), we propose a penalized version of $(\widehat{\mathcal{P}}_\lambda^{(m)})$ – which can also be seen as a perturbation of the atomic norm regularizer used in [7, 15]. We consider the following program:

$$\begin{aligned} \mathcal{R}_{\lambda,\rho} \in \operatorname*{argmin}_{u,z,\tau} \; & h(\mathcal{R}) + \frac{1}{2\rho}\|R - \sum_{|\boldsymbol{k}|\leq n} u_{\boldsymbol{k}}\Theta_{\boldsymbol{k}}\|^2 \\ \text{s.t.} \; & \mathcal{R} = \begin{bmatrix} R & z \\ z^* & \tau \end{bmatrix} \succeq 0, \quad R, \Theta_{\boldsymbol{k}} \in \mathbb{C}^{m^d \times m^d}. \end{aligned} \qquad (\widehat{\mathcal{P}}^{(m)}_{\lambda,\rho})$$

Obviously, the dual polynomial $\eta_{\lambda,\rho}$ deriving from $(\widehat{\mathcal{P}}^{(m)}_{\lambda,\rho})$ (using relation (2) which still holds for the penalized problem) may differ from η_λ, in particular with regard to their roots, thus compromising exact recovery, see Fig. 4. However, following an approach similar to [16], under some mild non-degeneracy hypothesis, it is possible to show that, for small enough values of ρ, $\mathcal{R}_{\lambda,\rho}$ is sufficiently close to $\mathcal{R}_\lambda$ to allow accurate support reconstruction (in particular they have the same rank, see Fig. 3). Numerical observations confirm that this regime exists.

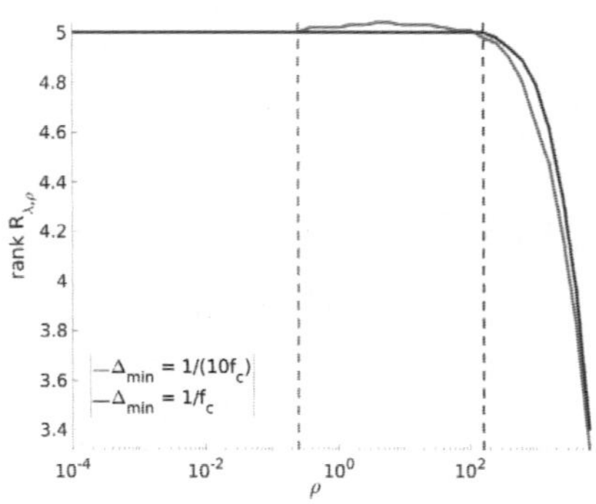

Figure 3: $f_c = 13$, $\lambda = 0.05$, $\sigma = 0$, $r = 5$ spikes. Evolution of the rank of $\mathcal{R}_{\lambda,\rho}$ w.r.t. parameter ρ for a serie of randomized signals with minimal separation distance Δ_{min}.

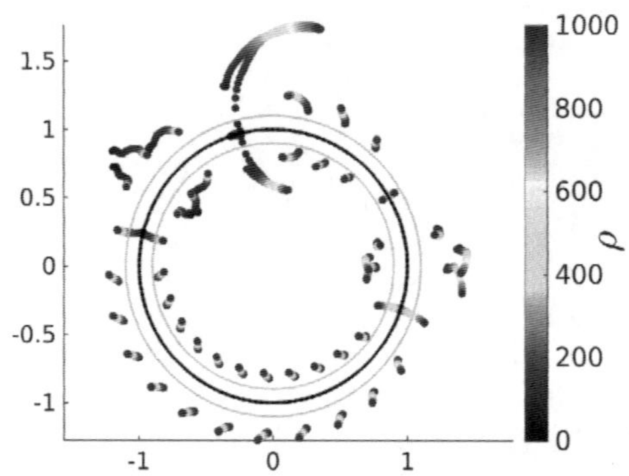

Figure 4: Same framework. Roots of $1-|\eta_{\lambda,\rho}|^2$ w.r.t. ρ. When $\rho = 0$, support is recovered from double roots, located on the unit circle. When $\rho > 0$, these may split, but remain identifiable when ρ is small.

4. Algorithm

In this section, we take advantage of the low-rank property of the solutions as well as of the convolutive structure of the Toeplitz constraint to build a robust and efficient numerical scheme for solving $(\widehat{\mathcal{P}}^{(m)}_{\lambda,\rho})$.

In the following, we note

$$f(\mathcal{R}) = u_0 + \tau + \frac{1}{2}\|\frac{y}{\lambda} + 2z\|^2 + \frac{1}{2\rho}\|R - \sum_{\|\boldsymbol{k}\|_\infty < m} u_{\boldsymbol{k}}\Theta_{\boldsymbol{k}}\|^2.$$

4.1. Superresolution algorithm

We propose a Frank-Wolfe scheme [17] to minimize f over the cone of positive semidefinite matrices. In order to return to optimization over a compact, we set a bound D_0 on the trace of the iterates such that $\operatorname{tr}\mathcal{R}_{\lambda,\rho} \leq D_0$. In practice, one may choose $D_0 = f(0)$.

To take advantage of the low-rank structure of the solutions, we store our iterates as $\mathcal{R} = ss^H$ (s^H being the conjugate transpose of s). We add a non-convex step similar to [6], consisting in a gradient descent on $F : s \mapsto f(ss^H)$, which can only improve the speed of convergence of the algorithm. We used a BFGS descent in our implementation, see Algorithm 1

IOP Conf. Series: Journal of Physics: Conf. Series **904** (2017) 012015 doi:10.1088/1742-6596/904/1/012015

Algorithm 1 Recovering dual polynomial

set: $s_0 = [0 \dots 0]^\top$, D_0 s.t. $\mathrm{tr}(s^\star) \leq D_0$ (bound on the domain)
while $\langle s_r s_r^H - v_r v_r^H, \nabla f(s_r s_r^H) \rangle \geq \epsilon$ **do**
1. *Compute next atom*: $v_r = D_0 \arg\min_{\|v\| \leq 1} v^\top \cdot \nabla f(s_r s_r^H) \cdot v$
2. *Update*: $\hat{s}_{r+1} = [\alpha_r s_r, \beta_r v_r]$ where $\alpha_r, \beta_r = \arg\min_{\alpha+\beta \leq 1} f(\alpha s_r s_r^H + \beta v_r v_r^H)$
3. *Corrective step*: $s_{r+1} = \mathbf{BFGS}(s \mapsto f(ss^H) : s \in \mathbb{C}^{(n+1)\times(r+1)}$, start from $\hat{s}_{r+1})$

end while
return $p_\lambda = \frac{y}{\lambda} + 2z$, where $\begin{bmatrix} z \\ \tau \end{bmatrix}$ is the last column of $s_{N+1} s_{N+1}^H$.

Both the positions and amplitudes of the solution may then be recovered from the coefficients p_λ. Algorithm 2 give the main steps of the reconstruction process. To recover the roots of $\eta_{\lambda,\rho}$ which correspond to the support of the solution μ_λ, we fix a tolerance around the unit circle in which they should be located, for small enough values of ρ, see Fig. 4.

Algorithm 2 Reconstruction via root-finding

input: p_λ coefficients of the dual polynomial
set $\epsilon > 0$, $R = \mathrm{roots}(p_\lambda)$
- $R1 \leftarrow \{r \in R\,;\, |1 - |r|| < \epsilon\}$ # *extract candidate roots*
- *positions*: $x \leftarrow \frac{1}{2\pi} \arg(R1)$ # *positions correspond to angles (remove symmetric roots)*
- *amplitudes*: $a \leftarrow (\Phi_x^* \Phi_x)^{-1} \Phi_x^* (y - \lambda p_\lambda)$, where $\Phi_x = \Phi|_x$ # *optimality relation*

return x, a

4.2. FFT-based computations
Step 1. in Algorithm 1 simply amounts to computing a leading eigenvector of $\nabla f(s_i s_i^H)$, using power iterations, which is based on matrix-vector products only. Given the particular form of f in our case, the sole costly operation will consist in

(i) computing the projection of $s_r s_r^H$ on Toeplitz tensors,
(ii) multiply the latter with a vector.

Both steps can be done efficiently using only Fast Fourier Transforms: indeed, one can show that the projector in (i) has a simple factorization involving only padding, un-padding, FFT and inverse FFT operators; on the other hand, the Toeplitz-vector product of (ii) may astutely be written as a circulant-vector product, which can also be computed using FFT, noticing that circulant matrices are diagonal in Fourier basis.

5. Numerics
The results that are presented in this section were all obtained in 1-D settings.

Accurate reconstruction essentially relies on the ability to detect correct roots among all of the roots of $\eta_{\lambda,\rho}$. In our simulations, we set manually a confidence interval around the unit circle, see Algorithm 2 ($\epsilon = 0.1$ worked well in general, but may need to be tuned according to λ, w and ρ). We will investigate more reliable criteria and clustering algorithms in future works.

An important property of Algorithm 1 is that it has finite convergence, thanks to the non-convex descent step. Even more remarkably, convergence in k steps when k is the number of spikes in the solution is almost always observed for values of ρ that are not extreme, Fig 6. For small ρ, convergence is slower and less stable, and in particular the number of BFGS iterations needed to reach a given precision greatly increases, see Fig. 5.

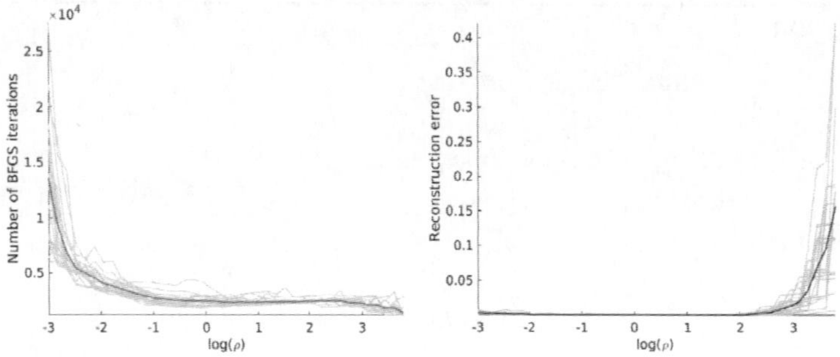

Figure 5: ($f_c = 13$, $\lambda = 0.05$, $\sigma = 0$, $r = 5$, $\Delta_{\min} = 1/f_c$). *Left*: total number of BFGS iterations (tolerance: 10^{-9} in terms of parameters changes; maximum number of iterations: 1000), summed over all Frank-Wolfe iterations. *Right*: $\|\mu_{\lambda,\rho} - \mu_\lambda\|_{\mathrm{BL}}^*$, where $\mu_{\lambda,\rho}$ and μ_λ are solutions of $(\widehat{\mathcal{P}}_{\lambda,\rho}^{(m)})$ (computed with Algorithms 1,2) and $(\widehat{\mathcal{P}}_{\lambda}^{(m)})$ (computed with CVX) respectively, and $\|\cdot\|_{\mathrm{BL}}^*$ is the dual bounded Lipschitz norm [18].

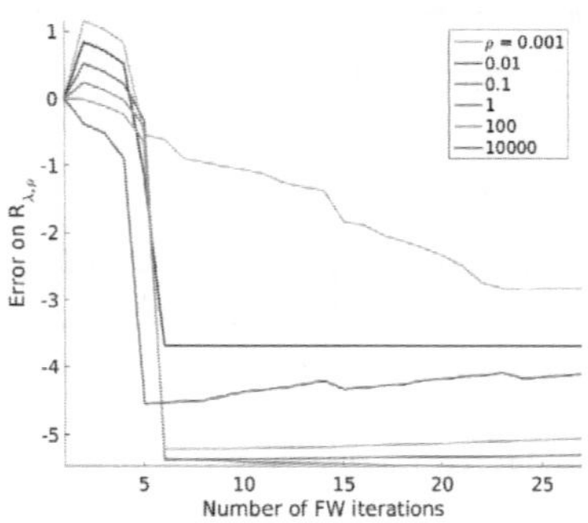

Figure 6: ($f_c = 13$, $\lambda = 0.05$, $\sigma = 0$, $r = 5$, $\Delta_{\min} = 1/f_c$). BFGS tolerance: 10^{-9}, maximum number of iterations: 1000. We measure $\|\mathcal{R}_{\lambda,\rho} - \mathcal{R}_\lambda\|$ (in logarithmic scale), $\mathcal{R}_\lambda$ being computed with CVX. We see that for reasonable values of ρ, convergence occurs after 5 steps.

Acknowledgments
This work was supported by grants from Région Ile-de-France.

References

[1] de Castro Y and Gamboa F 2012 *Journal of Mathematical analysis and applications* **395** 336–354
[2] Bredies K and Pikkarainen H K 2013 *ESAIM: Control, Optimization and Calculus of Variations* **19** 190–218
[3] Candès E and Fernandez-Granda C 2014 *Communications on pure and applied Mathematics* **67** 906–956
[4] Duval V and Peyré G 2017 *Inverse Problems* **33**
[5] Chen S S, Donoho D L and Saunders M A 2001 *SIAM Rev.* **43** 129–159
[6] Boyd N, Schiebinger G and Recht B 2015 The alternating descent conditional gradient method for sparse inverse problems *CAMSAP* pp 57–60
[7] Tang G, Bhaskar B N, Shah P and Recht B 2013 *IEEE Trans. Information Theory* **59** 7465–7490
[8] de Castro Y, Gamboa F, Henrion D and Lasserre J 2017 *IEEE Trans. Information Theory* **63** 621–630
[9] Kunis S, Peter T, Römer T and Von der Ohe U 2016 *Linear Algebra and its Applications* **490** 31–47
[10] Liao W and Fannjiang A 2014 *CoRR*
[11] Wei X and Dragotti P 2016 *IEEE Trans. Image Processing* **25**
[12] Lasserre J B 2001 *SIAM Journal on Optimization* **11** 796–817
[13] Dumitrescu B A 2007 *Positive trigonometric Polynomials and Signal Processing Applications* Signals and Communication Technology (Springer Publishing Company, Incorporated)
[14] Yang Z, Xie L and Stoica P 2016 *IEEE Trans. Information Theory* **62** 3685–3701
[15] Tang G, Bhaskar B and Recht B 2015 *IEEE Trans. Information Theory* **61** 499–512
[16] Vaiter S, Peyré G and Fadili J 2015 *Low Complexity Regularization of Linear Inverse Problems* (Springer International Publishing) pp 103–153
[17] Jaggi M 2013 Revisiting frank-wolfe: Projection-free sparse convex optimization *ICML* vol 28
[18] Dudley R 1966 *Studia Mathematica* **27** 251–268

IOP Conf. Series: Journal of Physics: Conf. Series **904** (2017) 012016 doi:10.1088/1742-6596/904/1/012016

Adaptive Order Non-Convex L_p-norm Regularization in Image Restoration

Manya Afonso and João Miguel Sanches

Institute of Systems and Robotics, Instituto Superior Técnico, Lisbon, Portugal

E-mail: mafonso,jmrs@isr.ist.utl.pt

Abstract. The optimization formulations of sparse signal recovery problems use a regularizer term which encourages the solution to be sparse or piece-wise smooth. Of special interest are the ℓ_p norms with $0 \leq p \leq 1$ which penalize large values and encourage the solution of the optimization problem to be sparse. In this work, we propose a new regularizer in which the exponent $p, 0 \leq p \leq 1$ of the ℓ_p-norm is a function of the signal element amplitude, leading to a function which is smooth and convex over non-negative values. This formulation adapts the value of the exponent p according to a sigmoid function applied on the signal values, to restrict its values to between 0 and 1. Experiments show that this formulation is more accurate in denoising, than the ℓ_1 and $\ell_{0.5}$ norm regularizers, and is more accurate than the ℓ_1-norm when used iteratively for deblurring or inpainting.

1. Introduction

Non-convex regularizers such as the ℓ_p norms with $0 \leq p < 1$ are of interest in sparse estimation, especially in problems such as compressive sensing because of their ability to recover sparser solutions from fewer measurements than the convex regularizers such as the ℓ_1 norm [3, 11]. If $\mathbf{x}$ is the image to be estimated, and $\mathbf{y}$ is a noisy observed image, both in vector representation, a widely used optimization formulation for estimating $\mathbf{x}$ is,

$$\min_{\mathbf{x}} \frac{1}{2}\|\mathbf{x}-\mathbf{y}\|_2^2 + \frac{\lambda}{2}\phi(\mathbf{x}), \tag{1}$$

where $\phi(.)$ is a non-convex regularizer, and $\lambda > 0$ is the regularization parameter. For the ℓ_p norm regularizer, it is defined as $\phi(\mathbf{x}) = (\sum_i |x_i|^p)^{1/p}$. Sometimes its pth power $\phi(\mathbf{x}) = (\ell_p(\mathbf{x}))^p = \sum_i |x_i|^p$ is used for mathematical tractability. If $\phi(.)$ is not convex, then eq (1) is not guaranteed to have a global minimum. For the ℓ_0 norm, the problem is combinatorial, therefore often a relaxation with the convex ℓ_1 norm is used, which has a closed form minimizer for (1), given by the soft threshold [7]. This is also called the LASSO problem [10].

The iteratively reweighted least squares (IRLS) method [3, 6] solves (1) by expressing a non-convex ℓ_p norm regularizer as a weighted quadratic norm, and therefore iteratively solving a least squares problem. A similar approach was used in [11] for a larger set of non-convex regularizers, by expressing (1) as a weighted LASSO problem.

For convex regularizers, a widely used solver is the split augmented Lagrangian and shrinkage algorithm (SALSA) [1], which belongs to the alternating direction method of multipliers (ADMM) family of algorithms, for which convergence conditions have been defined and proven

[8]. More recently, it has been shown empirically that ADMM can still converge even if the regularizer $\phi(.)$ is non-convex [15, 4]. It was proven in [14] that unlike the more general augmented Lagrangian method, ADMM can converge with a non-convex and possibly non-smooth objective function, as long as the objective can be separated into a smooth component and a possibly non-smooth and non-convex component, and the constraints are linear.

In [13], the authors solve the square root LASSO problem with a weakly convex regularizer which is convex over blocks of variables, and is defined in terms of the difference between the absolute value and a weighted square. In this paper, we follow a similar logic to define a non-convex regularizer which is convex over some intervals of interest. We propose a solver for problem (1) with this regularizer, based on IRLS. Experimental results show that this regularizer leads to a more accurate denoised image than using the ℓ_1 and $\ell_{0.5}$ norm regularizers, and that it is more accurate when used with SALSA for image deblurring and inpainting.

2. Proposed Approach

We define a new regularizer in which the exponent is not constant, but a function of the absolute amplitude,

$$\phi(\mathbf{x}) = \sum_i |x_i|^{q(|x_i|)}, \tag{2}$$

where the exponent q is a function of $|x|, x \in \mathbb{R}$. Since we are interested in values of q in the interval $[0, 1]$, a natural choice is the sigmoid function $\sigma(x) = 1/(1 + \exp(-x))$. This function is used in binary classification [10] because of it can be used as a smooth approximation to a non-linear thresholding, since it approaches 0 if $x < 0$ and 1 if $x > 0$. In our case, we are working with the absolute value $|x|$, therefore the threshold must be different from 0. We would also like to penalize different intervals of the dynamic range of the image, differently. We therefore adapt the sigmoid function to the following rule for the exponent q, $q(|x|) = 1/(1 + \exp(-\alpha(\log |x| - K)))$, where $\alpha > 0$ is a parameter controlling the level of smoothness and shape of the function $q(.)$, for values of α close to zero, the function varies slowly and for larger values, it switches between 0 and 1 with a steep slope, as shown in figure 1(a). The parameter $K \in \mathbb{R}$ is an additional parameter to control the threshold at which the switching occurs, and depends on the minimum value of $|x|$ below which it is considered negligible. In the limit that $|x|$ is very small, the term $\log |x|$ will be negative, leading to $q(|x|)$ tending to zero. In this case, in our formulation, we use the limit that any real value raised to an exponent tending to 0 tends to 1. By doing this, we have a regularizer which is a smooth function of x over $x \in (0, \infty)$. Note that this regularizer is not a norm.

For completeness of notation, we will define α as one of the parameters of $q(.)$, leading to

$$q(|x|, \alpha, q_0) = \frac{q_0}{1 + \exp(-\alpha(\log |x| - K))}, \tag{3}$$

and $\phi(\mathbf{x}, \alpha, q_0) = \sum_i |x_i|^{q(|x_i|, \alpha, q_0)}$, where q_0 is the maximum value of the exponent. In this paper, we will use $q_0 = 1$ unless stated otherwise.

It was experimentally found that for $\alpha > 0.5$, the function (2) is always convex over $x \geq 0$, in that its second order gradient is always positive. It can be seen in figures 1(b) and 1(c) that the regularizer $\phi(\mathbf{x})$ is convex for α equal to 0.51 and 1, but not for 0.1. By comparison, $\ell_{0.5}^{0.5}$ is clearly non-convex over the same range of values. The value of $\alpha = 0.51$ was also the one which was experimentally found to have the minimum value of $\phi(\mathbf{x})$ closest to 0, in this case 0.1, as marked by the blue circles in the plots in figure 1(c). To simplify notation, we will use $\phi(\mathbf{x})$ to indicate $\phi(\mathbf{x}, \alpha = 0.51, q_0 = 1)$ in the rest of this paper.

Solver for Denoising

We now solve the denoising problem by using our non-convex regularizer $\phi(\mathbf{x})$ in (1). Using the differentiation rule for functions of the form $f(x) = g(x)^{h(x)}$, the derivative of $\phi(\mathbf{x})$ with respect to an element x_i can be shown to be,

$$\frac{\partial \phi(\mathbf{x})}{\partial x_i} = \left(\frac{\partial q(|x_i|)}{\partial x_i} \log(|x_i|) + \frac{q(|x_i|)}{|x_i|} \right) \operatorname{sign}(x_i)|x_i|^{q(|x_i|)}. \tag{4}$$

Since the derivative of the sigmoid function is $d\sigma/dx = \sigma(x)(1 - \sigma(x))$, it can be shown that

$$\frac{\partial q(|x_i|)}{\partial x_i} = \frac{\alpha q(|x_i|)}{x_i} \left(1 - \frac{q(|x_i|)}{q_0} \right). \tag{5}$$

Using eq (4), the partial derivative of eq (1) *wrt* x_i can be expressed as

$$\begin{aligned} x_i - y_i + \frac{\lambda}{2} \left(\frac{\partial q(|x_i|)}{\partial x_i} \log(|x_i|) + \frac{q(|x_i|)}{|x_i|} \right) \operatorname{sign}(x_i)|x_i|^{q(|x_i|)} &= 0 \\ x_i - y_i + \underbrace{\frac{\lambda}{2} \left(\frac{\partial q(|x_i|)}{\partial x_i} \log(|x_i|) + \frac{q(|x_i|)}{|x_i|} \right) |x_i|^{q(|x_i|)-1}}_{=a(x_i)=a_i} x_i &= 0 \end{aligned} \tag{6}$$

Using an approach similar to IRLS, we calculate the value of x_i at iteration $t+1$ as

$$x_i^{(t+1)} = \frac{y_i}{1 + 0.5\lambda a_i^{(t)}}, \tag{7}$$

where a_i is a function of x_i as formulated in eq (6), calculated using the value at the previous iteration $x_i^{(t)}$.

If $\mathbf{x}$ is the wavelet representation of an image, when applying the regularizer $\phi(.)$, we use the formulation,

$$\min_{\mathbf{x}} \frac{1}{2} ||\mathbf{x} - \mathbf{W}^T \mathbf{y}||_2^2 + \frac{\lambda}{2} \phi(\mathbf{x}), \tag{8}$$

where $\mathbf{W}^T$ is the forward wavelet transform.

The proposed algorithm can be used as a denoising operator [5] in solvers such as SALSA [1] for deconvolution and inpainting problems,

$$\min_{\mathbf{x}} \frac{1}{2} ||\mathbf{A}\mathbf{x} - \mathbf{y}||_2^2 + \frac{\lambda}{2} \phi(\mathbf{W}^T \mathbf{x}), \tag{9}$$

where $\mathbf{x}, \mathbf{y}$ represent the pixels, $\mathbf{A}$ is the linear operator representing the convolution or the element-wise binary mask representing partial observation of pixels. The analysis formulation [9] is used to apply the proposed regularizer $\phi(.)$ on the wavelet representation of $\mathbf{x}$. The variable splitting used in SALSA ensures that at each iteration, the denoising problem of the form (1) is separable.

3. Results

The proposed approach was compared for denoising against soft thresholding the wavelet coefficients (4 level, redundant Daubechies 4) [7] and IRLS for denoising with the $(\ell_{0.5})^{0.5}$ regularizer, on the 256 × 256 Lena and Cameraman images with added Gaussian noise (SNR 10 dB). All experiments were performed in Matlab on Windows 7 on an Intel i5 based system with 4 GB of RAM. The stopping criterion for the proposed method was the relative difference

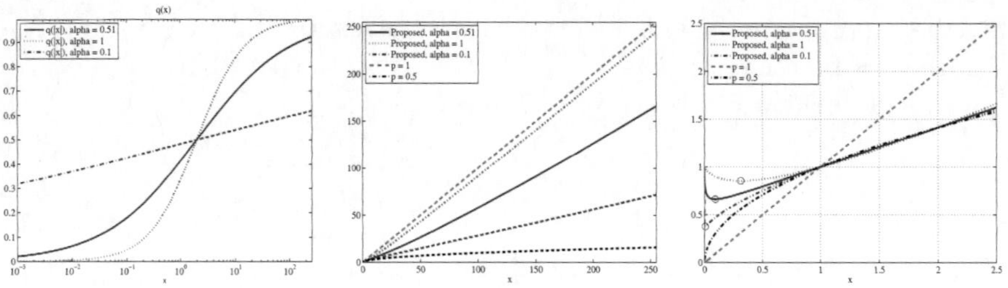

(a) (b) (c)

Figure 1. (left) Plot of sigmoid function for different values of α; (middle, right) plot of $\phi(.)$ for different values of α and other norms.

between successive iterations falling below 10^{-3}. The soft threshold and IRLS were both hand tuned to obtain the best set of parameters which produced the most accurate estimate. The figures of merit used are the Median Absolute Error (MAE), Improvement in Signal to Noise Ratio (ISNR), and the Structured Similiarity Index Measure [2].

We can see from table 1 that the proposed approach leads to a more accurate estimate than the soft threshold or IRLS with $(\ell_{0.5})^{0.5}$ regularizer, *i.e.*, it produces a larger ISNR and SSIM and lower MAE. The original and noisy images, along with the estimates obtained with the three methods are shown in figure 2.

For deblurring and inpainting, a few iterations of the proposed denoising solver is used as the denoising operator in SALSA, using the analysis wavelet prior formulation. The proposed approach is compared against SALSA with the ℓ_1 norm regularizer on the wavelet coefficients. The blur used is 9×9 uniform blur, and for inpainting, 40% of the pixels are randomly discarded. In both cases, Gaussian noise with SNR 40 dB is added. From table 1, we see that the proposed approach leads to a higher ISNR and SSIM and lower MAE than the ℓ_1 norm. The observed and estimated images are shown in figures 3 and 4. It can be seen from figures 4(b), 4(c), 4(e), and 4(f) that the proposed approach estimates better regions with edges such as the hair and hat edges in the Lena image, and the silhouette and tripod edges of the cameraman.

Table 1. Comparison of proposed approach with ℓ_1 and $\ell_{0.5}$ norm regularization, for image denoising, deblurring, and inpainting.

		Lena					Cameraman				
Problem	Method	MAE	ISNR (dB)	SSIM	Iters	Time (sec)	MAE	ISNR (dB)	SSIM	Iters	Time (sec)
Denoising SNR 10 dB	ℓ_1 (Soft)	7.139	2.654	0.675	-	0.01	8.928	2.960	0.542	-	0.01
	$\ell_{0.5}$ (IRLS)	5.339	4.113	0.760	12	2.00	6.741	4.407	0.634	13	2.27
	Proposed	5.006	**4.265**	**0.774**	12	4.85	6.286	**4.510**	**0.644**	13	5.64
Deblurring 9×9 uniform	ℓ_1 (SALSA)	2.525	6.454	0.873	45	8.76	2.283	6.489	0.860	61	11.27
	Proposed	2.510	**6.978**	**0.877**	7	12.56	2.186	**6.989**	**0.866**	7	12.67
Inpainting 40% missing	ℓ_1 (SALSA)	1.399	14.155	0.825	344	34.01	1.977	11.790	0.810	444	43.56
	Proposed	0.551	**15.747**	**0.873**	61	30.67	0.705	**12.652**	**0.849**	56	27.99

4. Conclusions and Future Work

In this paper, we have proposed a new regularizer for image restoration problems, which consists of an approximation to the ℓ_p norm, with the exponent p varying as a function of $|x|$. The

IOP Conf. Series: Journal of Physics: Conf. Series **904** (2017) 012016 doi:10.1088/1742-6596/904/1/012016

Figure 2. Denoising results with the Lena and Cameraman images: (a,f) original; (b,g) noisy (SNR 10 dB); Estimates with: (c,h) ℓ_1-norm (soft thresholding); (d,i) $\ell_{0.5}$-norm (IRLS); (e,j) proposed approach.

Figure 3. Deblurring results with the Lena and Cameraman images: (a,d) blurred (9×9 uniform blur, SNR 40 dB); Estimates with: (b,e) ℓ_1-norm; (c,f) proposed approach.

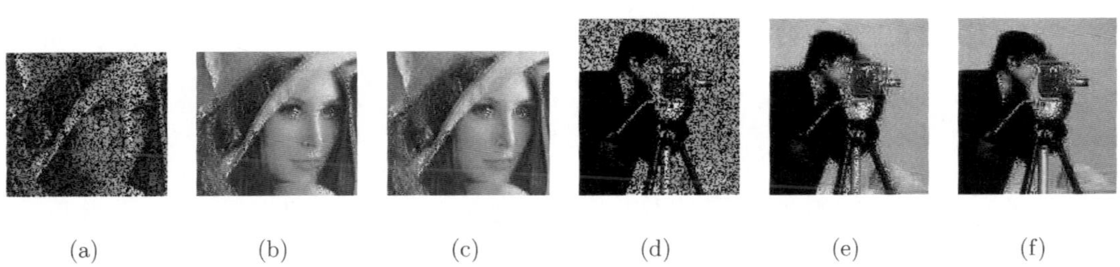

Figure 4. Inpainting results with the Lena and Cameraman images: (a,d) observed (40% pixels missing at random, SNR 40 dB); Estimates with: (b,e) ℓ_1-norm; (c,f) proposed approach.

resulting denoising problem is solved iteratively using IRLS. This formulation was experimentally found to lead to a more accurate estimate than the ℓ_1 or $\ell_{0.5}$ norm regularizers, for image denoising, deblurring, and inpainting with wavelets. Adjusting the exponent to get a convex function over both positive and negative values of x, and extending the formulation to total variation [12] will be addressed in a future paper.

Acknowledgements

This work was supported by the **Fundação para a Ciência e Tecnologia** (FCT), Portuguese Ministry of Science and Higher Education through a Post-doctoral fellowship (contract no. **SFRH/BPD/79011/2011**) and FCT project **UID/EEA/50009/2013**.

References

[1] M. Afonso, J. B. Dias, and M. Figueiredo, "Fast image recovery using variable splitting and constrained optimization," *IEEE Trans. on Im. Proc.*, vol. 19, no. 9, pp. 2345–2356, 2010.

[2] A. C. Bovik, *Handbook of image and video processing.* Academic Press, 2010.

[3] R. Chartrand and W. Yin, "Iteratively reweighted algorithms for compressive sensing," in *Acoustics, Speech and Signal Processing, 2008. ICASSP 2008. IEEE International Conference on.* IEEE, 2008, pp. 3869–3872.

[4] ——, "Nonconvex sparse regularization and splitting algorithms," in *Splitting Methods in Communication, Imaging, Science, and Engineering.* Springer, 2016, pp. 237–249.

[5] P. L. Combettes and V. R. Wajs, "Signal recovery by proximal forward-backward splitting," *Multiscale Modeling & Simulation*, vol. 4, no. 4, pp. 1168–1200, 2005.

[6] I. Daubechies, R. DeVore, M. Fornasier, and C. S. Güntürk, "Iteratively reweighted least squares minimization for sparse recovery," *Comm. on Pure and Applied Mathematics*, vol. 63, no. 1, pp. 1–38, 2010.

[7] D. Donoho, "De-noising by soft thresholding," *IEEE Trans. Information Theory*, vol. 41, no. 3, pp. 613–627, May 1995.

[8] J. Eckstein and D. Bertsekas, "On the DouglasRachford splitting method and the proximal point algorithm for maximal monotone operators," *Mathematical Programming*, vol. 55, no. 3, pp. 293–318, 1992.

[9] M. Elad, P. Milanfar, and R. Rubinstein, "Analysis versus synthesis in signal priors," *Inverse problems*, vol. 23, no. 3, p. 947, 2007.

[10] J. Friedman, T. Hastie, and R. Tibshirani, *The elements of statistical learning.* Springer series in statistics Springer, Berlin, 2001, vol. 1.

[11] G. Gasso, A. Rakotomamonjy, and S. Canu, "Recovering sparse signals with a certain family of nonconvex penalties and dc programming," *IEEE Transactions on Signal Processing*, vol. 57, no. 12, pp. 4686–4698, 2009.

[12] L. Rudin, S. Osher, and E. Fatemi, "Nonlinear total variation based noise removal algorithms," *Physica D*, vol. 60, pp. 259–268, 1992.

[13] X. Shen, L. Chen, Y. Gu, and H. So, "Square-root lasso with nonconvex regularization: An admm approach," *IEEE Signal Processing Letters*, vol. 23, no. 7, pp. 934–938, 2016.

[14] Y. Wang, W. Yin, and J. Zeng, "Global convergence of admm in nonconvex nonsmooth optimization," *arXiv preprint arXiv:1511.06324*, 2015.

[15] Z. Xu, S. De, M. Figueiredo, C. Studer, and T. Goldstein, "An empirical study of admm for nonconvex problems," *arXiv preprint arXiv:1612.03349*, 2016.

IOP Conf. Series: Journal of Physics: Conf. Series **904** (2017) 012017 doi:10.1088/1742-6596/904/1/012017

Real time groove characterization combining partial least squares and SVR strategies: application to eddy current testing

S Ahmed[1], M Salucci[2], R Miorelli[1], N Anselmi[2], G Oliveri[2], P Calmon[1], C Reboud[1] and A Massa[2,3]

[1]CEA LIST, Centre de Saclay, F- 91191 Gif-sur-Yvette, France
[2]ELEDIA Research Center (ELEDIA@UniTN - University of Trento)
Via Sommarive 9, I-38123 Trento, Italy
[3]ELEDIA Research Center (ELEDIA@L2S - UMR8506)
3 rue Joliot-Curie 91192 Gif-sur-Yvette, France
E-mail: andrea.massa@l2s.centralesupelec.fr

Abstract. A quasi real-time inversion strategy is presented for groove characterization of a conductive non-ferromagnetic tube structure by exploiting eddy current testing (ECT) signal. Inversion problem has been formulated by non-iterative Learning-by-Examples (LBE) strategy. Within the framework of LBE, an efficient training strategy has been adopted with the combination of feature extraction and a customized version of output space filling (OSF) adaptive sampling in order to get optimal training set during *offline* phase. Partial Least Squares (PLS) and Support Vector Regression (SVR) have been exploited for feature extraction and prediction technique respectively to have robust and accurate real time inversion during *online* phase.

1. Introduction

Nowadays non-invasive inspection through non-destructive testing and evaluation (NDT-NDE) is becoming very popular area of research for different applications. Among different electromagnetic NDT (E-NDT) methods, lower frequency eddy current testing (ECT) is widely used for the assessment of the integrity of the structure under test (SUT) by means of defect or crack detection, localization and characterization. Within this context, the necessity of quicker and reliable inverse solutions becomes the main priority in many industrial applications. In general, inversion strategy can be classified into iterative and non-iterative methods. Standard iterative approaches based on deterministic or stochastic algorithms [1]-[4] can be too cumbersome due to the solution of several hundred up to thousands forward problems in order to minimize a suitable cost function. This aspect can be computationally expensive and time demanding tasks. Alternatively, non-iterative techniques have recently shown to be able to address quite successfully inversion issues [5]-[6]. Nevertheless, due to the limitations on certain probe assessment, they cannot be applied directly to any kind of NDE problems. Among the non-iterative approaches, kernel-based strategies [7]-[8] have been presented in the past years while Learning-by-Examples (LBE) strategies have been proposed recently for quasi real time inverse solution [9]-[10]. Following the presented work at [9]-[10] LBE strategy can be formulated in two phases. At the preliminary *offline* phase, a fast and accurate model is built by generating a training set of input-output (I/O) pairs. During the second phase (*online* phase), the developed model from *offline* phase is used to predict the output associated to an unknown test sample. The performance of different LBE strategies depends on the particular choice of the training set generation in terms of samples selection in parameter space and features extraction in ECT signal space. Towards this end, this work describes an adaptive-sampling strategy which combines Partial Least Squares (PLS) [11] feature extraction and output space filling (OSF) sampling [8]. This strategy aims to uniformly explore the extracted feature space to have enough information for training set generation. Finally a set of Support Vector Regressors (SVR) [12] are exploited for accurate and quasi real-time inversion for groove characterization within a tube structure. Moreover, the robustness of the proposed *PLS-OSF/SVR* strategy is also evaluated for noisy test case based on numerical benchmark problem.

2. Mathematical formulation of forward and inverse problem

Let us consider a 2D axisymmetric configuration made by conductive tube having conductivity σ, relative permittivity $\varepsilon_r = 1$ and permeability $\mu_r = 1$ (Fig. 1). The tube is inspected by two coils which are excited by a time-harmonic current. The coils centred at $x = 0$ and $z = 0$ are working in differential mode and moving along the tube y axis. The tube is effected by a single axisymmetric groove that occupies a volumetric region Ω within the *SUT* (i.e., $\sigma(\mathbf{r}) \neq \sigma \quad \forall \mathbf{r} \in \Omega$ - Fig. 1). The impedance variation measured by the coils at the *k*-th scanning position with respect to the flawless scenario ($k = 1,\ldots,K$) is given by [13]

$$\Psi_k = -\frac{1}{I^2}\int_{\Omega} \mathbf{E}^{inc}(\mathbf{r} \mid \mathbf{r}_k) \cdot \boldsymbol{\rho}(\mathbf{r} \mid \mathbf{r}_k) d\mathbf{r} \tag{1}$$

I is the current flowing inside the coil while $\mathbf{E}^{inc}(\mathbf{r} \mid \mathbf{r}_k)$ is the incident field generated at position $\mathbf{r}$ in the unflawed tube ($\mathbf{r}_k = y_k$ represents the *k*-th coil position along the tube). $\boldsymbol{\rho}(\mathbf{r} \mid \mathbf{r}_k)$ represents the unknown induced current dipole density, which models the presence of the groove and is related to the total field, $\mathbf{E}^{tot}(\mathbf{r} \mid \mathbf{r}_k)$ that can be expressed by

$$\boldsymbol{\rho}(\mathbf{r} \mid \mathbf{r}_k) = [\sigma(\mathbf{r}) - \sigma]\mathbf{E}^{tot}(\mathbf{r} \mid \mathbf{r}_k). \tag{2}$$

Let us model the region Ω of the groove by a finite set of Q=3 descriptors $\underline{p} = (y_c, d_c, w_c)$. h_c and w_c are the height and width of the groove placing at y_c position along the tube axis with an angular extension of 360° (Fig. 1).

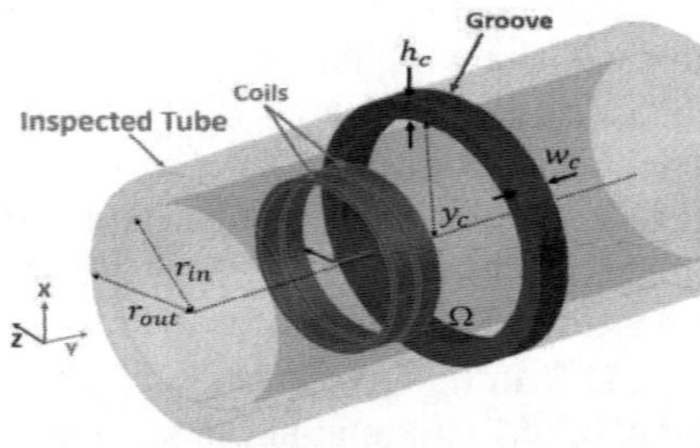

Figure 1. Tube geometry of the considered ECT problem.

Eq. (1) can be expressed in implicit form by means of the forward operator $\Phi\{.\}$ by $\Psi_k = \Phi\{\mathbf{r}_k, \underline{p}\}$, for $k = 1,\ldots,K$. CIVA [14] simulator is utilized for generating ECT signals associated to the coils differential impedance variation. The (unknown) inverse operator $\Phi^{-1}\{.\}$ allows to estimate the (unknown) groove parameters [i.e., $\underline{\tilde{p}} = (\tilde{p}_q; q = 1,\ldots,Q)$] by exploiting the information collected through ECT signals. Within the framework of LBE a regression problem is formulated for the estimation of $\Phi^{-1}\{.\}$ by suitably processing a set of N I/O pairs [i.e., a training set, $D_N = \left\{\left(\underline{\Psi}^{(n)}; \underline{p}^{(n)}\right); n = 1,\ldots,N\right\}$]. Due to the complex nature of the ECT signals, the n-th input is represented by the set $\underline{\Psi}^{(n)} = \left(\Re\{\Psi_k^{(n)}\}, \Im\{\Psi_k^{(n)}\}; k = 1,\ldots, K\right)$ of $F = 2K$ measured features associated to the n-th output $\underline{p}^{(n)} = \left(p_q^{(n)}; q = 1,\ldots,Q\right)$. The main goal of the proposed solution is to apply PLS feature extraction on F measured features to reduce the dimension of actual ECT feature space and perform adaptive sampling directly in the extracted feature space in order to retrieve the lowest N number of training samples during *offline* phase. This provides an exhaustive representation of the I/O relationship for optimal and almost real time inverse solution during *online* phase. The following steps describe the iterative procedure in order to adaptively select samples during the *offline* phase.

i. ***Initialization***- Generate $N = N_0$ number of initial samples by using a uniform grid sampling within the groove parameter space (i.e., $\underline{p}^{(n)}, n = 1,\dots, N$). $N_{\max}$ is the maximum desired/feasible training size (i.e., $N_0 < N_{\max}$). By using $\Phi\{.\}$ generate ECT coil signals and fill the $(N \times F)$ feature matrix $\underline{\underline{\Psi}}$ whose n-th row is represented by $\underline{\Psi}^{(n)}$. A matrix of groove parameters, $\underline{\underline{p}}$ having $(N \times Q)$ dimension is formed where $\underline{p}^{(n)}$ is the n-th row of $\underline{\underline{p}}$.

ii. ***Feature Extraction***- Build the $(N \times F)$ matrix $\underline{\underline{\Psi}}^T$ by subtracting each f-th column of $\underline{\underline{\Psi}}$ $(f = 1,\dots, F)$ from its mean value, μ_f and compute the $(N \times Q)$ matrix $\underline{\underline{p}}^T$ by subtracting each q-th column of $\underline{\underline{p}}$ $(q = 1,\dots, Q)$ from its mean value μ_p. Apply the *PLS* algorithm to linearly decompose $\underline{\underline{\Psi}}^T$ and $\underline{\underline{p}}^T$ as follows

$$\begin{aligned} \underline{\underline{\Psi}}^T &= \underline{\underline{T}} \times \underline{\underline{S}} + \underline{\underline{Y}} \\ \underline{\underline{p}}^T &= \underline{\underline{U}} \times \underline{\underline{Z}} + \underline{\underline{G}} \end{aligned} . \tag{3}$$

In Eq. (3) $\underline{\underline{T}} = \left(\underline{T}^{(n)}; n = 1,\dots,N\right)$ is the $(N \times J)$ matrix of Ψ-scores [$\underline{T}^{(n)} = \left(T_j^{(n)}; j = 1,\dots,J\right)$]. It is obtained from $\underline{\underline{\Psi}}^T$ through the $(F \times J)$ weight matrix $\underline{\underline{W}}$ (i.e., $\underline{\underline{T}} = \underline{\underline{\Psi}}^T \times \underline{\underline{W}}$). J is the number of extracted features ($J < F$). $\underline{\underline{Y}}$ and $\underline{\underline{G}}$ contain the $(N \times F)$ and $(N \times Q)$ residuals of the linear decomposition while $\underline{\underline{S}}$ and $\underline{\underline{Z}}$ are the $(J \times F)$ and $(J \times Q)$ matrices of loadings. The decomposition in Eq. (3) is aimed at maximizing the covariance between the corresponding columns of $\underline{\underline{T}}$ [i.e., $\left(\overline{T}_j; j = 1,\dots,J\right)$] and of the $(N \times J)$ matrix of p-scores $\underline{\underline{U}}$ [i.e., $\left(\overline{U}_j; j = 1,\dots,J\right)$]. This guarantees all the information about the ECT signal embedded inside $\underline{\underline{\Psi}}^T$ (i.e., inside $\underline{\underline{\Psi}}$) is compressed into $\underline{\underline{T}}$. Then an initial training set $\hat{D}_N = \left\{\left(\underline{T}^{(n)}; \underline{p}^{(n)}\right); n = 1,\dots, N\right\}$ is formed.

iii. ***Adaptive Sampling***- By using Latin Hypercube Sampling (*LHS*) select V candidate samples within the parameter space by $\underline{p}_{cand}^{(v)} = \left(p_{cand,q}^{(v)}; q = 1,\dots, Q\right), v = 1,\dots, V$. An estimation of the J-dimensional set of extracted features corresponding to each v-th candidate, $\tilde{\underline{T}}_{cand}^{(v)}$ is retrieved by applying a multi-dimensional linear interpolator on $\hat{D}_N$. Select the candidate sample $v = v^*$ through maximizing the minimum distance with the $\underline{T}^{(n)}$ sets in $\hat{D}_N$ [i.e., $\tilde{\underline{T}}_{cand}^{(v^*)} = \arg\left(\max_v \left\{\min_n d_{vn}\right\}\right)$]. d_{vn} is the Euclidean distance between $\tilde{\underline{T}}_{cand}^{(v)}$ ($v = 1,\dots, V$) and $\underline{T}^{(n)}$ ($n = 1,\dots, N$) [i.e., $d_{vn} = \sqrt{\sum_{j=1}^{J} \left(\tilde{T}_{cand,j}^{(v)} - T_j^{(n)}\right)^2}$]. The set of F measured features associated to the selected sample $\underline{\Psi}_{cand}^{(v^*)} = \left(\Re\left\{\Psi_{cand,k}^{(v^*)}\right\} \Im\left\{\Psi_{cand,k}^{(v^*)}\right\} k = 1,\dots,K\right)$ is computed by utilizing $\Phi\{.\}$. Finally, the set of extracted features is obtained by

$$\underline{T}_{cand}^{(v^*)} = \left(\underline{\Psi}_{cand}^{(v^*)}\right)^T \times \underline{\underline{W}} \tag{4}$$

where $\left(\underline{\Psi}_{cand}^{(v*)}\right)^T$ is obtained by subtracting each f-th element of $\underline{\Psi}_{cand}^{(v*)}$ ($f = 1,..., F$) from its mean value μ_f. Finally, updated the training set with $\hat{D}_{N+1} = \hat{D}_N \cup \left\{\underline{T}_{cand}^{(v*)}; \underline{p}_{cand}^{(v*)}\right\}$ and update $N = N+1$. This is also known as OSF sampling (i.e., candidate parameters are chosen such that features are uniformly distributed in the feature space).

iv. ***Stop Criterion-*** Stop adding new training samples for $N = N_{max}$. Otherwise, repeat from Step iii.

An ε-*SVR* is trained for each *q*-th parameter ($q = 1,...,Q$) of the groove, by exploiting the corresponding *q*-th set of I/O pairs $\hat{D}_{N,q} = \left\{\left(\underline{T}^{(n)}; p_q^{(n)}\right); n = 1,..., N\right\}$ on the generated training set. Moreover, a test set $\underline{\Psi}_{test}$ of *F* measured features associated to a previously-unseen groove parameter configuration $\underline{p}_{test}$ is projected through $\underline{\underline{W}}$ into the *J*-dimensional *PLS*-extracted features space [i.e., $\underline{T}_{test} = \underline{\Psi}_{test}^T \times \underline{\underline{W}}$. $\underline{\Psi}_{test}^T$ is obtained by subtracting each f-th element of $\underline{\Psi}_{test}$ ($f = 1,..., F$) from its mean value μ_f]. Finally, $\underline{T}_{test}$ is given as input to the *q*-th *SVR* in order to estimate the *q*-th parameter of the groove, $\tilde{p}_{test,q}$ for $q = 1,...,Q$.

3. Numerical validation

Let us consider a cylindrical metal tube (Fig. 1) of height by $(r_{out} - r_{in}) = 1.27$ mm and $\sigma = 1.0$ MS/m. Two axial current coils are working at 100 kHz in differential mode. The tube is effected by an external full groove ($\sigma(\mathbf{r}) = 0$ S/m, $\mathbf{r} \in \Omega$) located at $y_c = 23.75$ mm with fixed angular extension 360°. A fast metamodel is used to compute ECT signals of the coils over $K = K_x \times K_y = 1 \times 73 = 73$ probing locations with 0.2 mm spacing on *y* axis. The groove has variable height and width within the range $h_c \in [0.0635,\ 1.061]$ mm, $w_c \in [1.0,\ 3.0]$ mm respectively (i.e., $Q = 2$). $N_0 = 9$ samples are used to initialize the sampling loop while *J=4* features are extracted from $F = 2K = 2 \times 73 = 146$ actual measured features. $V = 100$ candidate samples are generated for each iterative step until $N = N_{max} = 49$ is obtained. Figure 2(*a*) shows the location of the sampling points in the of feature space (for graphical illustration, the resultant extracted feature space is shown for *J=2*). In Fig. 2(*b*), the sampling points have been mapped into groove parameter space for N_0 initial and the successive $(N_{max} - N_0)$ samples.

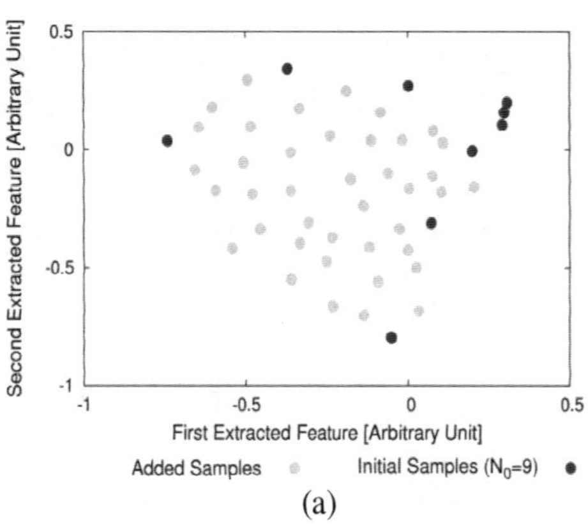

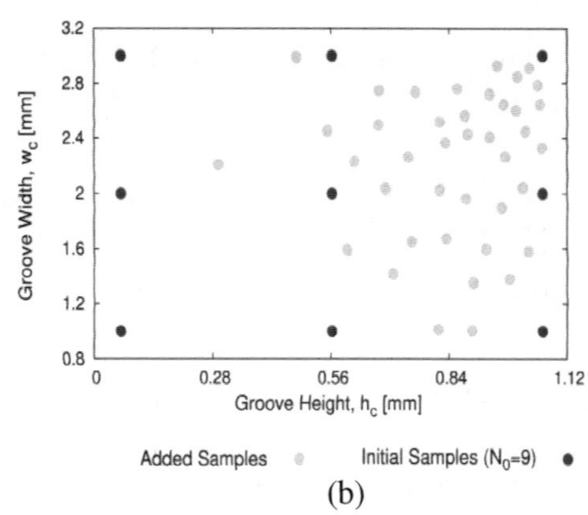

Figure 2. Training samples locations mapped on (*a*) the feature space and (*b*) the parameter space for *J*=2, N_0=9, N_{max}=49.

A set of $M = 1000$ previously-unseen test configurations (ECT signals) associated to randomly selected groove parameters has been generated. The robustness of the *PLS-OSF/SVR* schema is evaluated on the M test samples in presence of Additive White Gaussian Noise (AWGN). AWGN has been imposed by $SNR = [10, 20, 30, 40]$ dB according the definition mentioned in [10]. To access the performance on the prediction, normalized mean error, $NME_{p_q} = \frac{1}{M}\left(\sum_{m=1}^{M} | p_{test,q}^{(m)} - \tilde{p}_{test,q}^{(m)} | / | p_{test,q}^{(m)} |\right)$ is utilized where $p_{test,q}^{(m)}$ and $\tilde{p}_{test,q}^{m}$ are the actual and predicted q-th parameter of the m-th test sample ($m = 1,..., M$) respectively. It is evident from Figure 3, NME is slightly decreasing for h_c estimation with the increment of N and is not effected so much in presence of noise. Even at lower N value and noisy test set (e.g., $SNR = 20$ dB), h_c can be estimated. However, with increasing training samples additively, NME is decreasing significantly for w_c estimation for both noiseless and noisy test cases. That means, groove width (w_c) estimation suffers a lot for lower N and in presence of noisy test set. Consequently, more samples need to be added so as to represent enough impedance variation information due to the variation of w_c. This also helps the learning algorithm (i.e., SVR) to build optimal model for inverse solution. Therefore, the proposed approach uniformly explores the feature space by most significant features (i.e., J=4), and fills the parameter space mostly with higher values of h_c and uniform spacing of w_c [Fig. 2(*b*)].

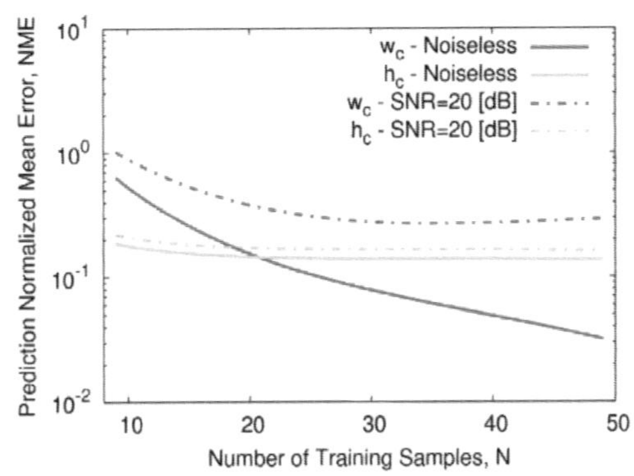

Figure 3. Prediction accuracy in terms of Normalized Mean Error (NME) vs. training size N for noiseless and noisy test set ($SNR = 20$ dB) for J = 4; Q = 2 by *PLS-OSF/SVR*.

In Figure 4, the robustness of *PLS-OSF/SVR* approach on noisy test set for estimating groove parameters is also depicted in terms of actual vs. predicted plots considering a fixed number of training samples (i.e., $N = 49$) and $J = 4$. With increasing noise (i.e., decreasing SNR value) on test set, h_c estimation [Fig. 4(a) - Fig. 4(d)] is not effected so much compare to w_c estimation [Fig. 4(e) - Fig. 4(h)]. However, even at noisy environment for a training set of $N = 49$ samples, w_c estimation shows reasonable accuracy (Fig. 3 and Fig. 4) by applying *PLS-OSF/SVR* technique.

The performance comparison between *PLS-OSF/SVR* technique for $J = 4$ and a standard *GRID/SVR* [10] approach is shown in Figure 5 in terms of the actual vs. predicted groove height and width considering the number of training samples $N = 49$. The considered training set for *GRID-SVR* is obtained by a full grid made by 7×7 (i.e., $N = 49$) samples having $F = 2K = 2\times73 = 146$ measured features. *PLS-OSF/SVR* shows better prediction accuracy than *GRID/SVR* by showing lower variance among the actual and predicted values (Fig. 5). Moreover, *GRID/SVR* fails to predict h_c and w_c with constant prediction in case of noisy test set that are not shown here (for sake of brevity). Contrarily, the transformation of actual F features into different extracted feature space by PLS makes *PLS-OSF/SVR* strategy more robust on noisy test set (Fig. 3 and Fig. 4).

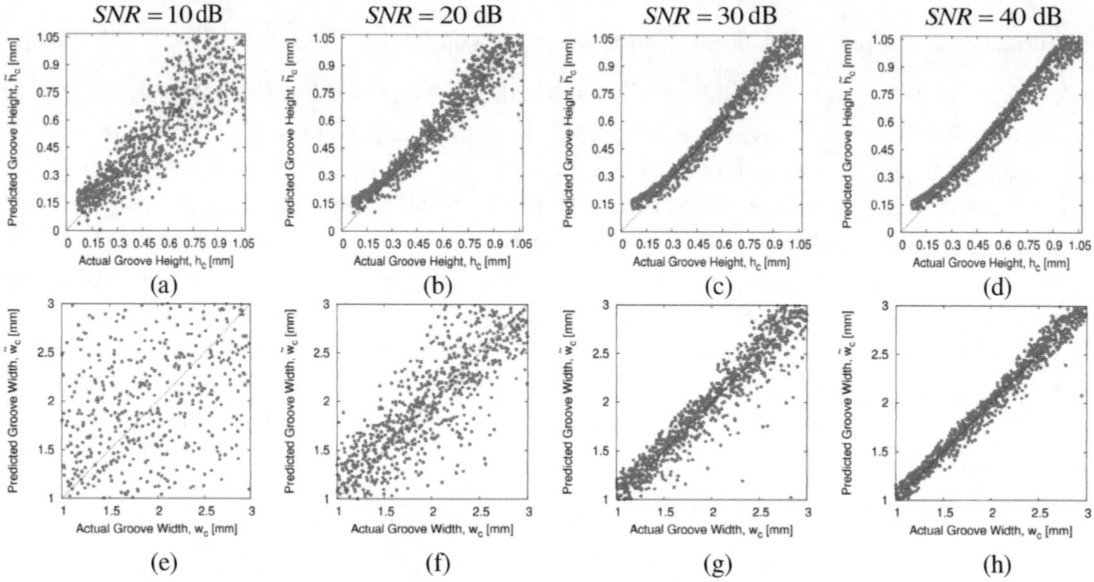

Figure 4. Actual vs. predicted plots for M=1000 test configurations at $SNR = [10, 20, 30, 40]$ dB, when N=49, $J = 4$ for groove (a) - (d) height, h_c and (e) - (h) width, w_c by *PLS-OSF/SVR*.

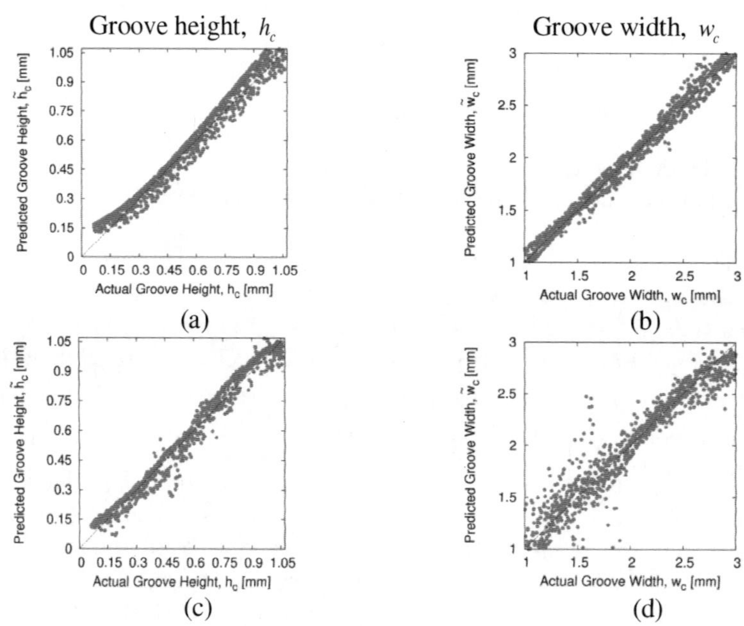

Figure 5. Actual vs. predicted h_c, w_c for M=1000 test configurations at *Noiseless* when N=49 for (a) - (b) *PLS-OSF/SVR* and (c) - (d) standard *GRID/SVR*.

4. Conclusion

In this work, we have shown an innovative real time inversion solution within the framework of Learning-by-Examples (LBE) for groove characterization in tube structure. PLS feature extraction combined with OSF sampling has shown the ability to build a more suitable training set than standard approach (i.e., full grid). The estimation accuracy and robustness of *PLS-OSF/SVR* strategy is numerically validated in presence of noisy test set and compared with standard *GRID/SVR* approach on noiseless test set. On the other hand, SVR shows the ability to deal with noisy test set if suitable training set is provided. Moreover, quasi real time inversion has been obtained for testing M=1000 test samples for 0.08s during *offline* phase on a standard laptop.

References

[1] Salucci M, Poli L and Massa A 2017 Advanced multi-frequency GPR data processing for non-linear deterministic imaging *Signal Processing- Special issue on Advanced Ground-Penetrating Radar Signal-Processing Techniques* **132** 306-318

[2] Salucci M, Oliveri G and Massa A 2015 GPR Prospecting through an inverse scattering frequency-hoping multi-focusing approach *IEEE Trans. Geosci. Remote Sens.* **53** 6573-6592

[3] Rocca P, Benedetti M, Donelli M, Franceschini D and Massa A 2009 Evolutionary optimization as applied to inverse problems *Inverse Problems* **25** 1-41

[4] Salucci M, Poli L, Anselmi N and Massa A 2017 Multifrequency particle swarm optimization for enhanced multi-resolution GPR microwave imaging *IEEE Trans. Geosci. Remote Sens.* **55** 1305-1317

[5] Henriksson T, Lambert M and Lesselier D 2011 Non-Iterative MUSIC-Type Algorithm for Eddy-Current Nondestructive Evaluation of Metal Plates *Studies in Applied Electromagnetics and Mechanics: Electromagnetic Nondestructive Evaluation (XIV)* **35** 22-29

[6] Tamburrino A,Vento A,Ventre S and Maffucci A 2016 Monotonicity imaging method for flaw detection in aeronautical applications *Studies in Applied Electromagnetics and Mechanics: Electromagnetic Nondestructive Evaluation (XIX)* **41** 284-292

[7] Douvenot R, Lambert M and Lesselier D 2011 Adaptive metamodels for crack characterization in eddy-current testing *IEEE Trans. Magn.* **47** 746-755

[8] Bilicz S, Lambert M, Gyimóthy S and Pávó J 2012 Solution of Inverse Problems in Nondestructive Testing by a Kriging-Based Surrogate Model *IEEE Trans. Magn.* **48** 495-498

[9] Salucci M, Ahmed S and Massa A 2016 An Adaptive Learning-by-Examples Strategy for Efficient Eddy Current Testing of Conductive Structures *Proc. Conf. EuCAP* **10**

[10] Salucci M, Anselmi N, Oliveri G, Calmon P, Miorelli R, Reboud C and Massa A 2016 Real-Time NDT-NDE Through an Innovative Adaptive Partial Least Squares SVR Inversion Approach *IEEE Trans. Geosci. Remote Sens.* **54** 6818-6832

[11] Wold S, Sjostrom M and Eriksson L 2001 PLS-regression: a basic tool of chemometrics *Chemometrics and Intelligent Laboratory Systems* **58** 109-130

[12] Smola A J and Scholkopf B A tutorial on support vector regression 2004 *Statistics and Computing* **14** 199-222

[13] Lesselier D and Razek A 2002 A Eddy current scattering and inverse scattering: Green's integral and variational formulations *Scattering: Scattering and Inverse Scattering in Pure and Applied Science. Part I. Scattering of Waves by Macroscopic Targets* 486-507

[14] CIVA: Simulation and Analysis for NDT. http://www.civa.cea.fr [Online]

Acknowledgment

This work has been partially supported by the SIRENA project (2014-2017) funded by DIGITEO (France) under the "Call for Chairs 2014" and French project ANR-ByPASS.

IOP Conf. Series: Journal of Physics: Conf. Series **904** (2017) 012018 doi:10.1088/1742-6596/904/1/012018

Multi-frequency direct sampling method in inverse scattering problem

Sangwoo Kang[1], Marc Lambert[1], and Won-Kwang Park[2]
[1]Group of Electrical Engineering, Paris (GeePs),UMR CNRS 8507, CentraleSupélec, Univ. Paris Sud, Université. Paris Saclay, UPMC Univ Paris 06, 3 & 11 rue Joliot-Curie 91192, Gif-sur-Yvette, France.
[2]Department of Information Security, Cryptology, and Mathematics, Kookmin University, Seoul, 02707, Korea

E-mail: sangwoo.kang@geeps.centralesupelec.fr, parkwk@kookmin.ac.kr and marc.lambert@geeps.centralesupelec.fr

Abstract. We consider the direct sampling method (DSM) for the two-dimensional inverse scattering problem. Although DSM is fast, stable, and effective, some phenomena remain unexplained by the existing results. We show that the imaging function of the direct sampling method can be expressed by a Bessel function of order zero. We also clarify the previously unexplained imaging phenomena and suggest multi-frequency DSM to overcome traditional DSM. Our method is evaluated in simulation studies using both single and multiple frequencies.

1. Introduction

This study considers the imaging of two-dimensional electromagnetic inclusions within a homogeneous space. The imaging is performed on measured scattered field data. Various non-iterative imaging algorithms have already been developed for this purpose, such as MUltiple SIgnal Classification [1, 2], the linear sampling method [3, 4], and Kirchhoff and subspace migrations [5, 6]. Although these methods are very effective for imaging unknown targets, they usually yield poor results when the number of incident waves is limited to one or only a few.

To overcome this difficulty, researchers have investigated the direct sampling method (DSM). According to [7, 8, 9], DSM has some merits: (i) it requires only one or a few incident waves for imaging the shapes and locations of unknown inclusions, (ii) it performs without any matrix operation (such as singular value decomposition), and (iii) it is highly tolerant to noise. However, these advantages are tempered by several drawbacks: (i) the direction of propagation is crucial for identifying the shapes and locations of inclusions, (ii) there are many artifacts in the map of DSM, and (iii) an inclusion with significantly smaller size or permittivity than other inclusions is difficult to identify by DSM.

To overcome these limitations, we improve the traditional DSM by applying multiple frequencies. The effectiveness of the technique is demonstrated on various synthetic and experimental data. The results confirm that the suggested imaging technique identifies unknown targets with dielectric contrast against the background space and improves the traditional DSM.

The remainder of the study is structured as follows. Section 2 introduces the two-dimensional direct scattering problem in the presence of small dielectric inhomogeneities. In section 3, we

IOP Conf. Series: Journal of Physics: Conf. Series **904** (2017) 012018 doi:10.1088/1742-6596/904/1/012018

briefly review single-frequency DSM, show that DSM can be expressed by a Bessel function of order zero, and propose our multi-frequency-based DSM. Section 4 presents the numerical and experimental results of both single and multiple frequency evaluations, and section 5 provides a short conclusion.

2. Two-dimensional direct scattering problem

In this section, we consider two-dimensional direct scattering from small dielectric inclusions τ_m, represented as $\tau_m = \mathbf{y}_m + \alpha_m \mathbf{B}_m$, $m = 1, 2, \cdots, M$. Let τ be the set of these scatterings; i.e., $\tau = \bigcup_m \tau_m$. For simplicity, we assume that all τ_m are balls with radius α_m located at $\mathbf{y}_m$. We also assume that $\mathbf{B}_m$ is a simply connected domain (circle). Let ω be the angular frequency and ε_m and μ_m be the dielectric permittivity and magnetic permeability of τ_m, respectively. ε_0 and μ_0 are defined analogously. In this paper, we consider only the permittivity contrast case, i.e., we set $\mu_m \equiv \mu_0 = 1$. Then we can define the piecewise constant permittivity $\varepsilon(\mathbf{x})$ as follows:

$$\varepsilon(\mathbf{x}) = \begin{cases} \varepsilon_m, & \text{if } \mathbf{x} \in \tau_m \\ \varepsilon_0, & \text{if } \mathbf{x} \in \mathbb{R}^2 \backslash \overline{\tau}. \end{cases} \tag{1}$$

For simplicity, we set $\varepsilon_0 = 1$. The wavenumber k is then given by $k = \omega\sqrt{\varepsilon\mu} = 2\pi/\lambda$, where λ is the wavelength. We assume that $\alpha_m \ll \lambda$ for all m and $k|\mathbf{y}_m - \mathbf{y}_{m'}| \gg 0.25$ for $m \neq m'$. This means that the τ_ms are well-separated from each other.

Let $u(\mathbf{x}; k)$ be the time-harmonic total field satisfying the following Helmholtz equation:

$$\Delta u(\mathbf{x}; k) + \omega^2 \varepsilon(\mathbf{x}) u(\mathbf{x}; k) = 0 \quad \text{in} \quad \mathbb{R}^2 \backslash \overline{\tau}, \tag{2}$$

with transmission conditions on the boundaries $\partial\tau_m$ for all m. As is well-known, the total field $u(\mathbf{x}; k)$ can be decomposed into an incident field $u^i(\mathbf{x}; k)$ and a scattered field $u^s(\mathbf{x}; k)$ such that $u(\mathbf{x}; k) = u^i(\mathbf{x}; k) + u^s(\mathbf{x}; k)$. In this study, we consider plane-wave illumination, i.e., we let $u^i(\mathbf{x}; k) = \mathrm{e}^{ik\mathbf{x}\cdot\mathbf{d}}$ be the incident field with propagation direction $\mathbf{d} \in \mathbb{S}^1$. Here, $\mathbb{S}^1$ denotes the two-dimensional unit circle. Note that $u^s(\mathbf{x}; k)$ satisfies the Sommerfeld radiation condition

$$\lim_{\mathbf{r}\to\infty} \left(\frac{\partial u^s(\mathbf{x}; k)}{\partial \mathbf{r}} - iku^s(\mathbf{x}; k) \right) = 0$$

uniformly in all directions $\mathbf{r} = |\mathbf{x}| \longrightarrow +\infty$. We assume that $u^s(\mathbf{x}_j; k)$, $j = 1, 2, \cdots, J$, is measured on a simply connected curve Γ. For simplicity, we set Γ as a circle with a large radius a such that, for $\mathbf{x} \in \Gamma$ and $m = 1, 2, \cdots, M$, $k|\mathbf{y}_m - \mathbf{x}| \gg 0.25$.

Based on small volume expansion(See [10]), $u^s(\mathbf{x}; k)$ can be expressed as the following asymptotic expansion formula for $j = 1, 2, \cdots, J$:

$$u^s(\mathbf{x}_j) = \sum_{m=1}^{M} \alpha_m^2 (\varepsilon_m - \varepsilon_0) |\mathbf{B}_m| \mathrm{e}^{ik\mathbf{d}\cdot\mathbf{y}_m} \Phi(\mathbf{x}_j, \mathbf{y}_m) + O(\alpha_m^2), \tag{3}$$

where $\Phi(\mathbf{x}, \mathbf{y})$ is the two-dimensional fundamental solution

$$\Phi(\mathbf{x}, \mathbf{z}) = \frac{\mathrm{i}}{4} H_0^1(k|\mathbf{x} - \mathbf{z}|). \tag{4}$$

3. Single- and multi-frequency direct sampling method

In this section, we briefly survey the DSM developed in [7], introduce the representative formula of the DSM imaging function, and propose a multi-frequency DSM to improve the imaging performance. The imaging function of the DSM is defined as

$$\mathcal{I}_{\mathrm{SF}}(\mathbf{z}; k) = \frac{|\langle u^s(\mathbf{x}_j; k), \Phi(\mathbf{x}_j, \mathbf{z}) \rangle_{L^2(\Gamma)}|}{\|u^s(\mathbf{x}_j; k)\|_{L^2(\Gamma)} \|\Phi(\mathbf{x}_j, \mathbf{z})\|_{L^2(\Gamma)}}, \tag{5}$$

where

$$\langle a(\mathbf{x}_j), b(\mathbf{x}_j)\rangle_{L^2(\Gamma)} := \sum_{j=1}^{J} a(\mathbf{x}_j)\overline{b(\mathbf{x}_j)} = \int_\Gamma a(\mathbf{x})\overline{b(\mathbf{x})}d\mathbf{x} \quad \text{and} \quad \|a(\mathbf{x}_j)\| := \langle a(\mathbf{x}_j), a(\mathbf{x}_j)\rangle_{L^2(\Gamma)}.$$

Referring to [7], DSM postulates that $u^s(\mathbf{x})$ can be approximated as

$$u^s(\mathbf{x}) \approx \sum_{m=1}^{M} W_m \Phi(\mathbf{x}, \mathbf{y}_m),$$

where W_m denotes the weight corresponding to τ_m. Then the following relation

$$\int_\Gamma u^s(\mathbf{z})\overline{\Phi}(\mathbf{z}, \mathbf{x})\mathrm{d}S \approx \frac{1}{k}\sum_{m=1}^{M} W_m \mathrm{Im}\left(\Phi(\mathbf{y}_m, \mathbf{x})\right) \approx \frac{1}{k}\sum_{m=1}^{M} J_0(k|\mathbf{z}-\mathbf{y}_m|), \tag{6}$$

we can observe that

$$\mathcal{I}_{\mathrm{SF}}(\mathbf{z}; k) \propto \sum_{m=1}^{M} J_0(k|\mathbf{z}-\mathbf{y}_m|).$$

From the above, we can see that if $\mathbf{z} = \mathbf{y}_m \in \tau_m$, then $\mathcal{I}(\mathbf{z}) \approx 1$, whereas if $\mathbf{z} \notin \tau$, then $\mathcal{I}(\mathbf{z}) < 1$. Therefore, we can identify the locations of τ_m by finding the $\mathbf{y}$ such that $\mathcal{I}(\mathbf{z}) \approx 1$. This method seems to be entirely reasonable, but leaves some phenomena unexplained. For example, τ_m with relatively small size or permittivity are very difficult to locate by this method. Motivated by this fact, we derive the following representation of $\mathcal{I}_{\mathrm{SF}}(\mathbf{z}; k)$ where the maximum value comes from the Hölder inequality. A detailed derivation will appear in an extended version of this study.

$$\mathcal{I}_{\mathrm{SF}}(\mathbf{z}; k) = \left|\sum_{m=1}^{M} \alpha_m^2(\varepsilon_m - \varepsilon_0)|\mathbf{B}_m| J_0(k|\mathbf{z}-\mathbf{y}_m|)\right| \left(\max\left|\sum_{m=1}^{M} \alpha_m^2(\varepsilon_m - \varepsilon_0)|\mathbf{B}_m|\right|\right)^{-1}. \tag{7}$$

As evident in (7), a τ_m with smaller size or permittivity than the other τ_ms will be difficult to locate because $\mathcal{I}_{\mathrm{SF}}(\mathbf{z}; k)$ is maximized at the inclusion with largest size or permittivity.

To overcome this difficulty, we exploit the well-known fact that multiple frequencies guarantee better results than single-frequency applications [5, 6]. Therefore, we introduce the following multi-frequency DSM. For N-different wavenumbers k_n, $n = 1, \cdots, N$, the imaging function can be introduced as

$$\mathcal{I}_{\mathrm{MF}}(\mathbf{z}) = \frac{1}{N}\left|\sum_{n=1}^{N} \langle u^s(\mathbf{x}_j; k), \Phi(\mathbf{x}_j, \mathbf{z})\rangle_{L^2(\Gamma)}\right| \left(\max\left|\sum_{n=1}^{N} \langle u^s(\mathbf{x}_j; k), \Phi(\mathbf{x}_j, \mathbf{z})\rangle_{L^2(\Gamma)}\right|\right)^{-1} \tag{8}$$

This approach improves the imaging by DSM. A careful analysis of $\mathcal{I}_{\mathrm{MF}}(\mathbf{z})$ will appear in the extended study.

4. Numerical experiments

To test the proposed approach, we performed diverse numerical experiments on synthetic and experimental single- and multi-frequency data.

IOP Conf. Series: Journal of Physics: Conf. Series **904** (2017) 012018 doi:10.1088/1742-6596/904/1/012018

4.1. Synthetic data

In the numerical experiments on synthetic data, we consider the imaging of small dielectric inclusions in a homogeneous space by DSM. We explored two cases; varying the properties (size and permittivity) of the inclusions and maintaining the same properties (size and permittivity) of the inclusions.

We assumed that there exist three small dielectric inclusions $\tau = \bigcup_m \tau_m \subset \mathbb{R}^2$, $m = 1, 2, 3$ located at $\mathbf{y}_1 = (0.3, -0.3)$, $\mathbf{y}_2 = (-0.4, -0.2)$, and $\mathbf{y}_3 = (-0.3, 0.5)$. $J = 30$ receivers are placed at $\mathbf{x}_j \in \Gamma$, where Γ is a circle with radius 5 centered at the origin. We assume a single propagation direction, $\mathbf{d} = (1, 1)/\sqrt{2}$. In the single-frequency application, we set $k = 2\pi/\lambda$ with $\lambda = 0.4$; in the multiple-frequency application, we set $k_n = 2\pi/\lambda_n$, where $\lambda_1 = 0.7$ and $\lambda_N = 0.4$ with $N = 10$. Here, λ_n are uniformly distributed in the interval $[0.4, 0.7]$.

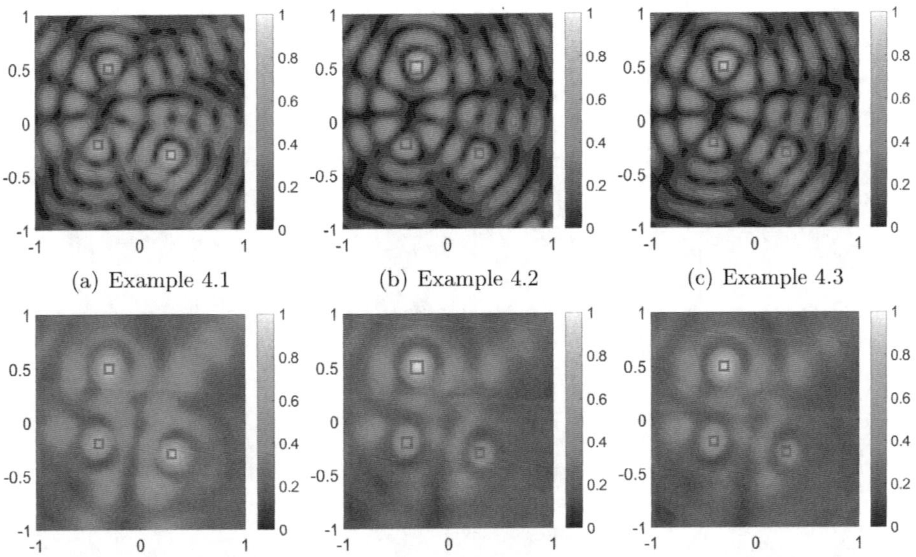

(a) Example 4.1 (b) Example 4.2 (c) Example 4.3

(d) Example 4.1 (e) Example 4.2 (f) Example 4.3

Figure 1. Imaging results of synthetic data: Single-frequency DSM (Top line) and multi-frequency DSM (Bottom line)

Example 4.1 (Case of $\alpha_m = \alpha_{m'}$ and $\varepsilon_m = \varepsilon_{m'}$) *We first set $\alpha_m = 0.1\lambda$ and $\varepsilon_m = 3$ for all $m = 1, 2, 3$. As shown in Figure 1(a), the locations of all inclusions are certificated with high accuracy. In (7), we confirmed that the value of $\mathcal{I}_{\mathrm{SF}}(\mathbf{z}; k)$ depends on the permittivities and sizes of the inclusions. As the permittivities and radii of all inclusions are identical in this case, $\mathcal{I}_{\mathrm{SF}}(\mathbf{z}; k)$ is maximized at $\mathbf{y}_m$, $m = 1, 2, 3$. As expected, the application of multiple frequencies reduces the artifacts; therefore, all inclusions are easily identified (see Figure 1(d)).*

Example 4.2 (Case of $\alpha_m \neq \alpha_{m'}$ and $\varepsilon_m = \varepsilon_{m'}$) *We then set $\alpha_1 = 0.1\lambda$, $\alpha_2 = 0.12\lambda$, and $\alpha_3 = 0.14\lambda$ and maintained the permittivity as $\varepsilon_m = 3$ for $m = 1, 2, 3$. In this case, the value of $\mathcal{I}_{\mathrm{SF}}(\mathbf{z}; k)$ depends only on the sizes of the inclusions. Because τ_3 is larger than the other inclusions, $\mathcal{I}_{\mathrm{SF}}(\mathbf{z}; k)$ is maximized at $\mathbf{y}_3 \in \tau_3$. Meanwhile, the values of $\mathcal{I}_{\mathrm{SF}}(\mathbf{z}; k)$ at $\mathbf{y}_1 \in \tau_1$ and $\mathbf{y}_2 \in \tau_2$ are smaller than that at $\mathbf{y}_3 \in \tau_3$. Therefore, the locations of τ_1 and τ_2 are not easily clarified (see Figure 1(b)). Fortunately, the locations of all inclusions are identifiable in the map of $\mathcal{I}_{\mathrm{MF}}(\mathbf{z})$ (Figure 1(e)).*

Example 4.3 (Case of $\alpha_m = \alpha_{m'}$ and $\varepsilon_m \neq \varepsilon_{m'}$) *In this case, we set $\varepsilon_1 = 3$, $\varepsilon_2 = 4$, and $\varepsilon_3 = 5$, maintaining the radius at $\alpha_m = 0.1\lambda$ for $m = 1, 2, 3$. The imaging result is exhibited in Figure 1(c). As all inclusions have the same radius, the value of $\mathcal{I}_{\text{SF}}(\mathbf{z}; k)$ depends only on the permittivities of the inclusions. This means that similar to the results in Example 4.2, the locations of τ_1 and τ_2 are not easily certificated. Fortunately, the multiple-frequency applications guarantee a good result in this case (see Figure 1(f)).*

4.2. Experimental data
Next, we applied DSM to a small U-shaped structure constructed from a metallic material. The experimental data of this structure were borrowed from [11]. We considered the cases of one transmitter (50°, 140°, 230°, and 320°), four transmitters, and a full set of transmitters (36 transmitters, 0° $\sim$ 360° with step size 10°). The single-frequency DSM was applied at 8 GHz and the multi-frequency DSM used seven frequencies ranging from 4 to 16 GHz with a step size of 2 GHz. The imaging results of the single- and multi-frequency applications are exhibited in Figures 2 and 3, respectively.

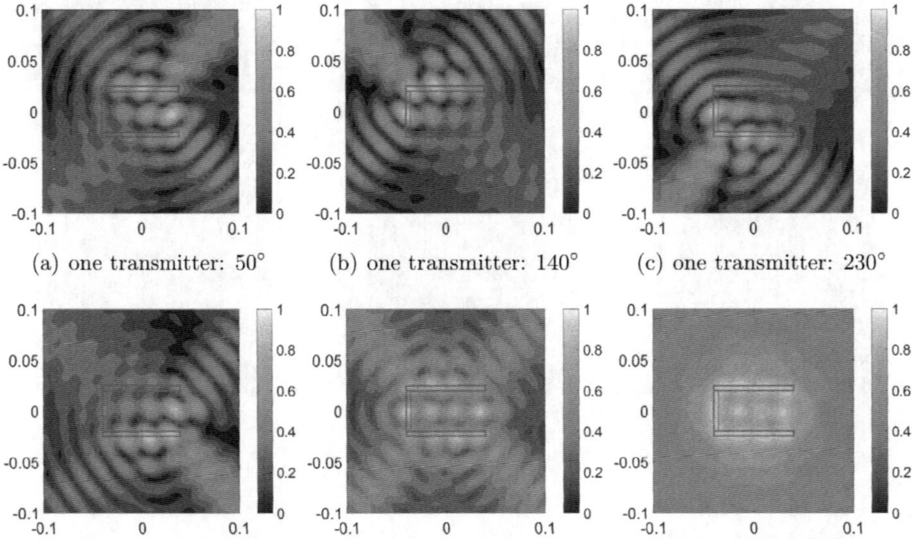

(a) one transmitter: 50° (b) one transmitter: 140° (c) one transmitter: 230°

(d) one transmitter: 320° (e) four transmitters (f) full transmitters

Figure 2. Imaging results of experimental data: Single-frequency DSM at 8 GHz

The location and shape of the inclusion are difficult to identify in the single transmitter case (see panels (a), (b), (c), and (d) of Figure 2). In the case of four transmitters, the shape is outlined but remains unclear (Figure 2(e)). In the full-transmitters case, the shape of the material is well clarified (Figure 2(f)). Unlike the results of the synthetic experiment, although our method reduces the artifacts, it does not fully recognize the shape of the material, as shown in Figure 3(a)–(e). Now, we compare the imaging from DSM with full transmitters and one from level set method with single-frequency(see (a) and (b) of Figure 4). Even though level set method is accurate, and efficient techniques, it requires more steps to construct shape like appropriate velocities for advancing the level set function.

5. Conclusion

In this study, we applied single- and multi-frequency DSMs to synthetic and experimental data. From the representative formula of the DSM imaging function and the results of

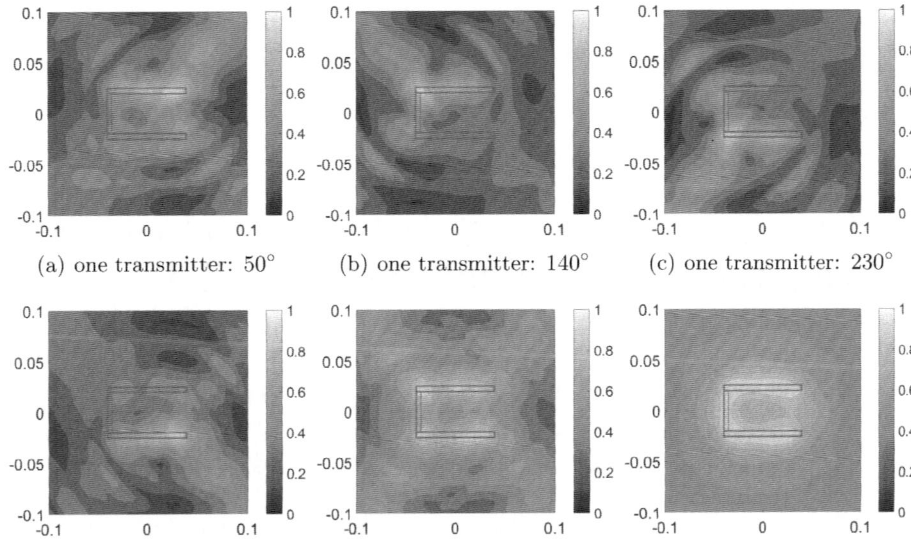

(a) one transmitter: 50° (b) one transmitter: 140° (c) one transmitter: 230°

(d) one transmitter: 320° (e) four transmitters (f) full transmitters

Figure 3. Imaging results of experimental data: Multi-frequency DSM

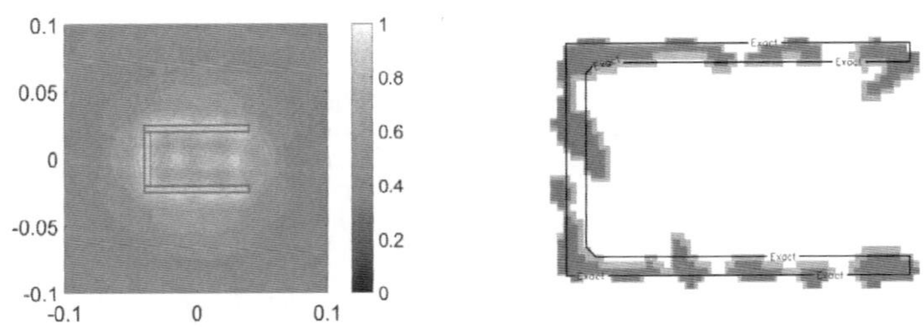

(a) Direct sampling method (b) Level set method(See [12])

Figure 4. Comparison between DSM and Level set method with experimental data at 8 GHz

numerical experiments, we confirmed that our DSM reasonably images synthetic data but requires improvement on experimental data. Identifying the cause of this non-ideal performance and improving our technique will be a forthcoming work. Recently, we found that the DSM is highly related to Kirchhoff migration. Establishing this relation would be an interesting project.

References

[1] Ammari H, Iakovleva E and Lesselier D 2006 *Multiscale Model. Sim.* **3** 597–628
[2] Ammari H, Iakovleva E, Lesselier D and Perrusson G 2007 *SIAM J. Sci. Comput.* **29** 674–709
[3] Cakoni F and Colton D 2003 *Inverse Prob.* **19** 279–295
[4] Colton D, Haddar H and Monk P 2002 *SIAM J. Sci. Comput.* **24** 719–731
[5] Ammari H, Garnier J, Kang H, Park W K and Sølna K 2011 *SIAM J. Appl. Math.* **71** 68–91
[6] Park W K 2015 *J. Comput. Phys.* **283** 52–80
[7] Ito K, Jin B and Zou J 2012 *Inverse Prob.* **28** 025003
[8] Ito K, Jin B and Zou J 2013 *Inverse Prob.* **29** 095018
[9] Li J and Zou Z 2013 *Inverse Probl. Imag.* **7** 757–775
[10] Ammari H and Kang H 2004 *Reconstruction of Small Inhomogeneities from Boundary Measurements* (*Lecture Notes in Mathematics* vol 1846) (Berlin: Springer-Verlag)
[11] Belkebir K and Saillard M 2001 *Inverse Prob.* **17** 1565–1571
[12] Ramananjaona C, Lambert M and Lesselier D 2001 *Inverse Problems* **17** 1585–1595

IOP Conf. Series: Journal of Physics: Conf. Series **904** (2017) 012019 doi:10.1088/1742-6596/904/1/012019

3D reconstruction of surface cracks using bi-frequency eddy current images and a direct semi-analytic model

Caifang Cai[1], Thierry Bore[2], Florentin Delaine[3], Nicolas Gasnier[3], Eric Vourc'h[3]

[1]L2S, CentraleSupelec/CNRS/Université Paris-sud/Université Paris-Saclay, France
[2]School of Civil Engineering, University of Queensland, St Lucia, Australia
[3]SATIE, ENS Paris-Saclay /Université Paris-Saclay /CNRS, France
E-mail: eric.vourch@satie.ens-cachan.fr, bore.thierry@gmail.com, caifang.cai@l2s.centralesupelec.fr, romain.soulat@thalesgroup.com

Abstract. We propose a method for reconstructing 3D surface cracks in metallic parts form eddy current (EC) images. To do so, we use a semi-analytic direct model for representing the interactions between EC and a crack. In order to cope with the inverse problem consisting in reconstructing 3D cracks, we propose the use of a genetic algorithm based on the computation of bi-frequency EC images. Numerical experiments are carried out in order to analyze the performance of the reconstruction method for different signal to noise ratios. Moreover, the results obtained with a bi-frequency genetic algorithm are compared to those obtained with a mono-frequency algorithm.

Introduction

The non destructive evaluation (NDE) is an important issue for the industries submitted to high security and reliability requirements, such as the nuclear and aeronautic industries. It is in particular necessary to make sure that the metallic parts of the plants or of the aircrafts do not feature corrosion or cracks, whether at the manufacturing step or during the use of the part. Ultrasounds [1] and eddy currents (EC) [2] are the most widely used NDE techniques. The latter, which we consider in this paper, has the advantage of being sensitive to the defects mentioned above, while being easy to implement and robust. Beyond the detection of defects, the evaluation of their features is necessary. Indeed, accurately evaluating defects dimensions makes it possible to envisage appropriate reparation and optimization of the maintenance operation. EC NDE is still focusing a lot of research works. The ill posed nature of the inverse problem being part of the issues to overcome [3-5]. Apart from carrying out precise defects reconstruction, it is also important to design inversion methods featuring a reasonable computation time.
The inversion of EC signals can be envisaged considering total variation regularization methods [6], fast interpolation models (like meta-models obtained from databases) and the estimation of a restricted number of parameters characterising the shape of the defects [7]. Since a main difficulty of the inversion relies on the establishment of a confidence criterion on the results of the algorithm, statistical approaches (such as Markov chain Monte Carlo [8]) also are considered. Of course, they require a high number of resolutions of the direct problem. Non-iterative algorithms also are proposed based on punctual tests of the inspected structure with near real-time NDE applications in view [9].

In order to deal with an NDE problem, it is useful combine an experimental set-up and a signal processing technique designed to match each other as well as the kind of problem that is considered (geometry of the inspected part, kind of defects...). Regarding EC techniques one can distinguish two main types of experimental approaches: those implementing either pulsed EC [10] or on harmonic EC [11].

In this paper, we consider the EC systems for which the excitation is harmonic and the current distribution is uniform [12, 13]. Indeed, such a configuration is sensitive to the presence of cracks and generates a significant magnetic field component perpendicular to the surface of the part. In addition, in order to dispose of information relating to the 3 dimensions (3D) of the inspected part, we consider multi-frequency imagery systems. Indeed, an image, which consists in a cartography of the magnetic field component normal to the surface of the part, provides with 2D information. On the other hand, given the skin effect, the measurement of images for different excitation frequencies facilitates the discrimination of the information relative to the depth of the part.

Moreover, it has been demonstrated that a semi-analytic modeling approach such as the distributed point source method (DPSM) enabled building a direct model suitable for representing the interactions between a crack and uniform EC flows [13]. This direct model featuring a rather fast computation time, the reconstruction of cracks by means of an iterative algorithm based on the computation of EC images can be envisaged [14]. Such an approach was previously carried out considering the reconstruction of 2D surface cracks from mono-frequency EC images [15].

In this paper, we propose the extension of this method to the reconstruction of 3D surface cracks based on the use of bi-frequency EC images.

This paper is organized as follows: in section 2, we describe the EC bi-frequency imaging configuration which we consider and we remind the main points of the direct semi-analytic model to be applied to this configuration. In section 3, we propose an approach for solving the inverse problem consisting in reconstructing 3D surface cracks by means of a genetic algorithm (GA) [16] operating in a bi-frequency configuration. In section 4, we analyze the estimation performances provided by this algorithm and we compare them to a similar approach using only mono-frequency images.

EC configuration and direct model

The EC configuration we consider corresponds to a uniform EC flow [12, 13] perpendicular to the largest surface of a crack present in a metallic part. Such a choice enables maximizing the interactions between the EC and the crack and subsequently generating a significant magnetic field distribution $\boldsymbol{b}_z$ perpendicularly to the surface of the part (Figure 1). Note that in practice, if the crack direction is not a priori known, one may try and maximize $\boldsymbol{b}_z$ by optimizing the orientation of the inducer generating the EC. As the EC distribution is subject to the skin effect, the EC intensity decreases exponentially as the depth z increases. Moreover, the skin depth δ, which represents an approximation of the penetration depth of the EC within the part, vary as the inverse of the root mean square of the excitation frequency f:

$$\delta = 1/\sqrt{\pi f \sigma \mu_0 \mu_r} \qquad (1)$$

where μ_0 is the magnetic permeability in a vacuum, μ_r is the relative magnetic permeability and σ is the electric conductivity of the medium.

Therefore, the measurement of EC images (that is of 2D $\boldsymbol{b}_z$ cartographies at the surface of the part) at different frequencies provides with information relating to the 3 dimensions of the part which can be useful with a view to the 3D reconstruction of cracks.

We have previously proposed a direct model suitable for the configurations described above [13-15]. $\boldsymbol{b}_z$ is computed as the sum of the contributions of sources radiating according to Green's functions. The sources, which are placed at the metal/air interface and within the crack volume, take both the skin effect and the boundary conditions into account. The model thus requires a discretization of the volume of the metallic part and of the interface. Note that for the sake of accuracy of the modeling it is suitable that the interface be meshed respecting a distance between the sources lower than a third of the characteristic distance of the electromagnetic problem, which is typically in the order of the skin depth δ. The model, which is based on the assumption that the presence of a crack represents a small perturbation with respect to a sound volume [17], has been previously validated by comparison to finite elements simulations and measurements [13].

Figure 1 provides with bi-frequency images computed with the model. They correspond to the case of an aeronautic aluminum alloy such that $\sigma = 17.6$ MS.m^{-1} and $\mu_r = 1$ and featuring a crack of width, length and depth W_x, L_y and D_z equal to 1 mm, 4 mm and 5 mm respectively.

These images correspond to f=100 Hz and f= 1200 Hz EC frequencies, which corresponds to 12.6 mm et 3.6 mm skin depths respectively. The main parameters of the model, such as those regarding the distretization of the workspace, are provided in Table 1. The same parameters will be used for every simulations reported in the paper.

Table 1. DPSM direct model configuration.

	x dimension		y dimension		z dimension	
	(mm)	(pixels)	(mm)	(pixels)	(mm)	(pixels)
Active sources planes (metal)	10	$n = 20$	10	$n = 20$	6	$p = 12$
b_z cartography	54	$m = 40$	54	$m = 40$		
Interface sources plane	@f= 100 Hz			@f= 1200 Hz		
	x and y dimensions			x and y dimensions		
	(mm)		(Pixels)	(mm)		(Pixels)
	≅ 4		$q = 13$	≅ 1		$q = 45$

Given the symmetry of the EC problem, the images feature two lobes and are symmetrical with respect to the x axis (Figure 1). It is to be noted that the higher frequency image, which carries information mostly regarding the top of the crack, features a lower spatial spread than the lower frequency image, which carries information regarding a deeper volume of the structure. According to this typical result, in an inversion algorithm the use of high frequency images should contribute to precisely estimate the parameters W_x, L_y , while that of low frequency images are due to improve the estimation of D_z.

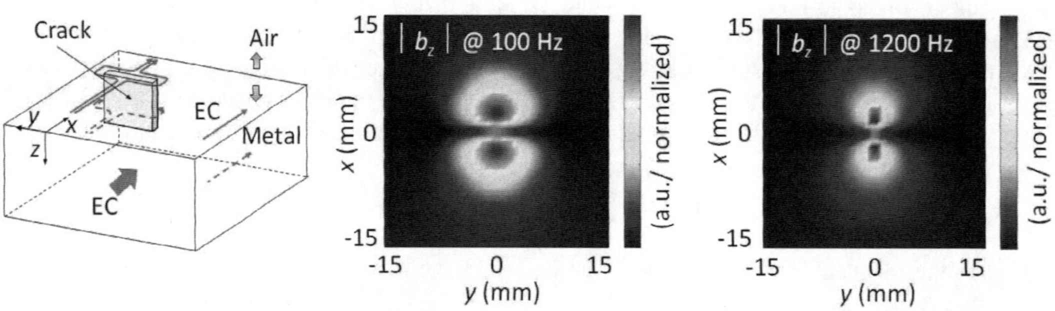

Figure 1. Interactions between a uniform EC flow and a 3D crack featuring a parallelepiped shape; synthetic $|\boldsymbol{b}_z|$ images at 100 Hz and 1200 Hz, for a metallic part featuring σ= 17.6 MS.m^{-1} and μ_r = 1 and for a crack such that $\{W_x = 1$ mm, $L_y = 4$ mm, $D_z = 5$ mm$\}$.

Inverse problem

The accuracy and computational efficiency of the direct model described in section 2 [13-15] led us to design an iterative crack reconstruction algorithm relying on the comparison of EC images computed with this model. In addition, according to the reasons given in section 2, with 3D crack reconstruction in view, the algorithm should use images at different frequencies.
Due to the complexity of the EC problem, different cracks are likely to provide with close EC images. Therefore, rather than an algorithm based on a gradient method, we chose to design a genetic algorithm (GA). Indeed, thanks to random variations of the features of the tested cracks at each iteration (mutation operation), the GA minimizes the risk to converge towards a local minimum.

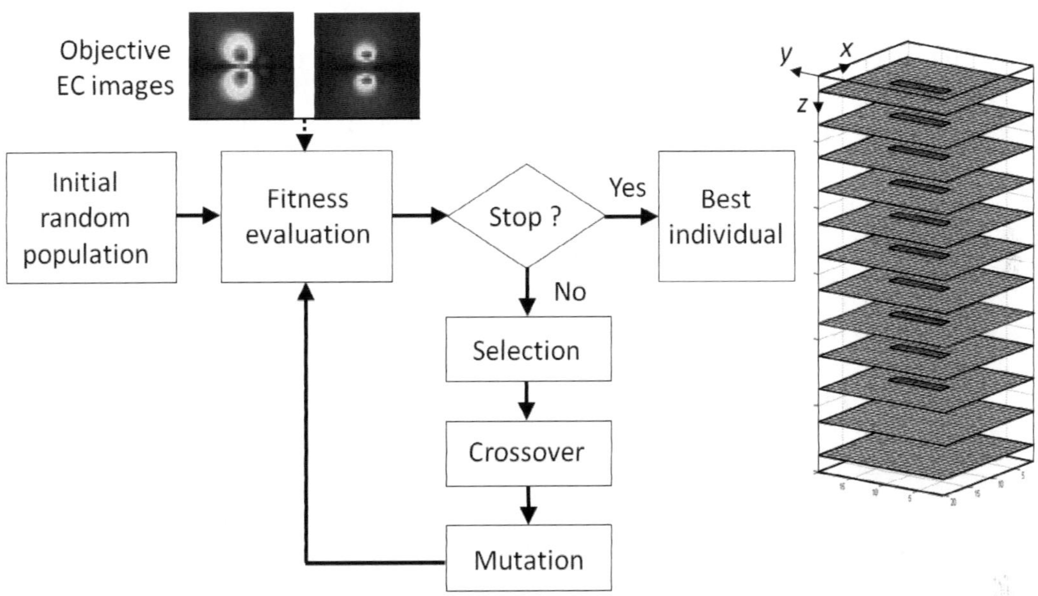

Figure 2. Genetic algorithm architecture for the reconstruction of 3D surface cracks from bi-frequency EC images.

The architecture of the GA is described in Figure 2. It uses a population of 8 individuals (cracks). The latter is randomly initialized while ensuring a minimum fitness of each individual. It is to ne noted that the population size was not optimized for minimizing the number of iterations, but simply chosen as 8. At each iteration, the fitting between bi-frequency EC images of the individuals and reference images corresponding to the researched crack is evaluated. Images comparison is carried out according to a normalized weighted relative square error ε_{wrse}, which gives all the more weight to a pixel that its energy is high. Let us denote $\boldsymbol{b_z} \in \mathbb{C}^M$ the vertical magnetic field at the surface of the metallic structure, where $M = m \times m$ is the image (i.e. cartography) dimension. ε_{wrse}, which is a relative error between $\boldsymbol{b_z}$ and a reference $\boldsymbol{b_{z0}}$ (output of the researched crack) is defined as follows:

$$\varepsilon_{wrse} = \frac{[\boldsymbol{b_z}-\boldsymbol{b_{z0}}]' W [\boldsymbol{b_z}-\boldsymbol{b_{z0}}]}{\boldsymbol{b_{z0}}' W \boldsymbol{b_{z0}}} \qquad (2)$$

where $W = diag|\boldsymbol{b_{z0}}|$ is the diagonal matrix supporting $|\boldsymbol{b_{z0}}|$ on the diagonal. In (2), the normalization is performed with respect to the total energy (or weighted energy) of the reference $\boldsymbol{b_{z0}}$.

The bi-frequency fitting criterion is based on the sum of the squares of the mono-frequency ε_{wrse} (such a sum proved more efficient than a product as regards the convergence of the algorithm).
Once evaluated, the population is modified in 3 steps: (i) selection (by means of a fortune wheel), (ii) crossover between the individuals and (iii) mutation: the 3 dimensions W_x, L_y et D_z of the individuals randomly mute according to given Gaussian laws.

Genetic algorithm performance analysis

GAs such as described in section 3 have been implemented in order to reconstruct cracks of dimensions such that {0.5 mm ≤ W_x ≤ 10 mm, 0.5 mm ≤ L_y ≤ 10 mm 0.5 mm ≤ D_z ≤ 6 mm}. The used EC frequencies are f =100 Hz and f = 1200 Hz. This choice is a trade off providing δ skin depths adapted to the considered cracks depths and to a meshing condition (the interface sources must be spaced out of a distance lower than δ/3) compatible with a fast computation time of the direct model.
Here we report on the performances of 3 different GAs: on the one hand 2 mono-frequency GAs using images at 100 Hz and 1200 Hz respectively, on the other hand a bi-frequency GA using images at both these frequencies. Numerical experiments have been carried out for 4 signal to noise ratios (SNR) comprised between 3 and 10 dB. 10 dB being a typical SNR of EC experimental systems.
Figures 4 and 5 compare the estimation performance of theses algorithms in the case of 2 cracks featuring the same width W_x = 1 mm and length L_y = 4 mm, but different D_z depths equal to 1 mm and 5 mm respectively (Figure 3). Each point of the convergence statistics reported in Figures 4 and 5 results from 30 drawings of the corresponding GA, the number of iterations being fixed at 30 per drawing which guaranties the GA to converge, according to our experience.

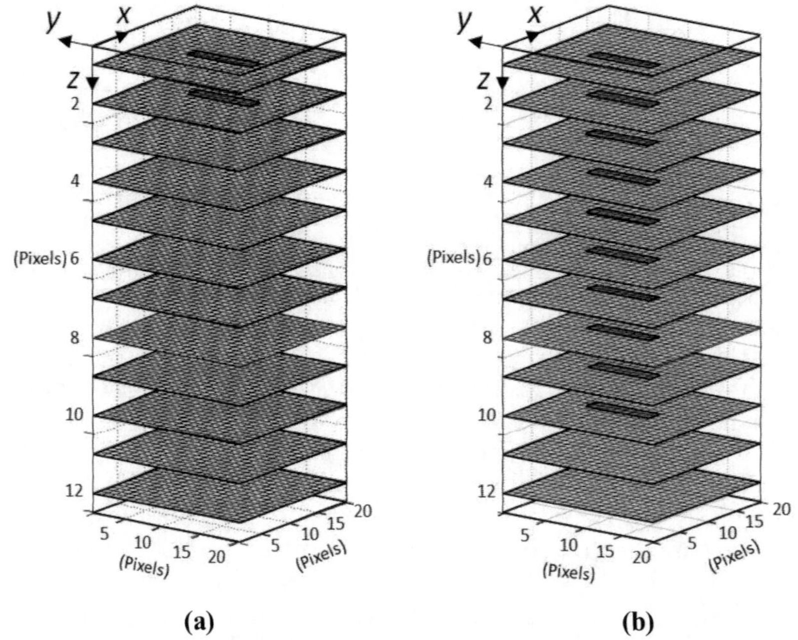

Figure 3. (a) Crack corresponding to the simulations reported in Figure 4 {W_x = 1 mm, L_y = 4 mm, D_z = 1 mm} ⇔ {W_x = 2 pixel, L_y = 8 pixels, D_z = 2 pixels}; (b) crack corresponding to the simulations reported in Figure 5 {W_x = 1 mm, L_y = 4 mm, D_z = 5 mm}⇔{W_x = 2 pixel, L_y = 8 pixels, D_z = 10 pixels}. The discretization of the DPSM direct model is as described in Table 1.

IOP Conf. Series: Journal of Physics: Conf. Series **904** (2017) 012019 doi:10.1088/1742-6596/904/1/012019

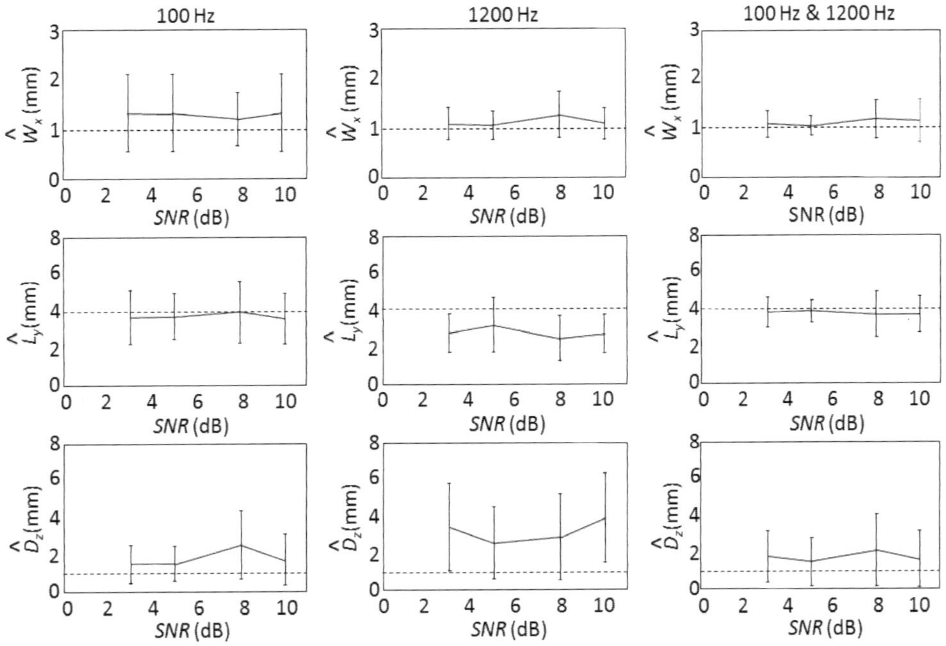

Figure 4. Expectancy and standard deviation of $\widehat{W_x}$ and $\widehat{L_y}$ *and* $\widehat{D_z}$ for 30 drawings of mono-frequency and bi-frequency GAs operating at 100 Hz and 1200 Hz with 30 iterations. Features of the researched crack: $\{W_x = 1\text{ mm}, L_y = 4\text{ mm}, \ D_z = 1\text{ mm}\} \Leftrightarrow \{W_x = 2\ \text{ pixel}, L_y = 8\ \text{ pixels}, D_z = 2\text{ pixels}\}$.

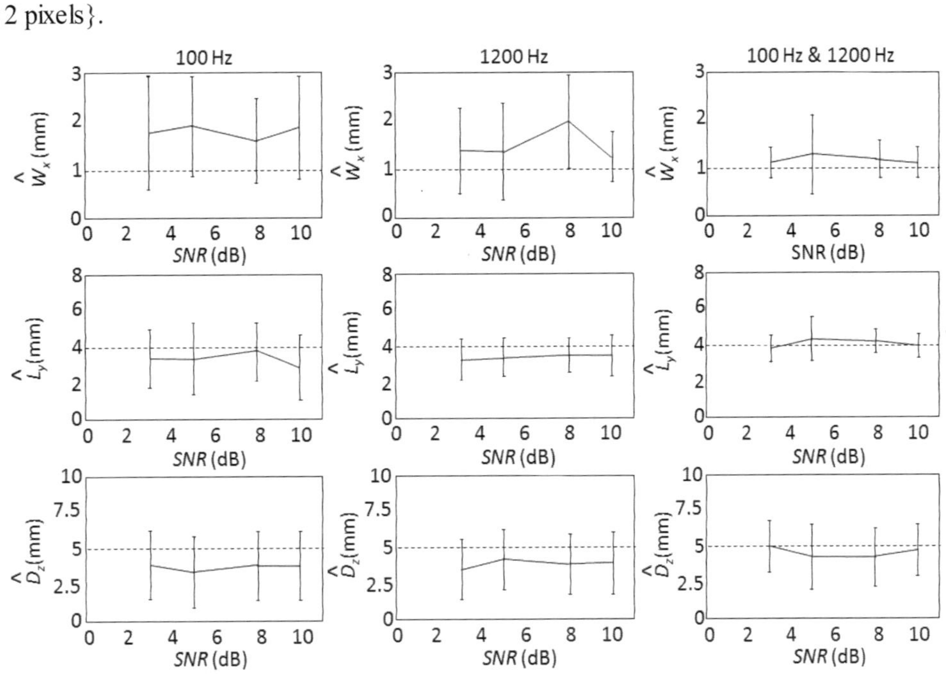

Figure 5. Expectancy and standard deviation of $\widehat{W_x}$ and $\widehat{L_y}$ *and* $\widehat{D_z}$ for 30 drawings of mono-frequency and bi-frequency GAs operating at 100 Hz and 1200 Hz with 30 iterations. Features of the researched crack: $\{W_x = 1\text{ mm}, L_y = 4\text{ mm}, \ D_z = 5\text{ mm}\} \Leftrightarrow \{W_x = 2\ \text{ pixel}, L_y = 8\ \text{ pixels}, D_z = 10\text{ pixels}\}$.

As regards of the results reported in Figure 4, it is to be noted that the mono-frequency GA at 100 Hz is the one providing the less accurate $\widehat{W_x}$ estimate, with an average bias lower than 0.5 mm (1 pixel) and a standard deviation in the order of 1 mm. The mono-frequency GA at 1200 Hz provides with poor accuracy of $\widehat{D_z}$, with a bias and a standard deviations of about 2 mm, whatever the SNR. Such results can be attributed to the spatial spread of low frequency images and to the small skin depth of EC at higher frequency, as mentioned in section 2.
As far as the bi-frequency GA is concerned, precise estimation of the three dimensions of the crack is carried out, whatever the SNR.
Regarding the reconstruction of the deeper crack (Figure 5), the mono-frequency GA at 100 Hz once again lacks accuracy in the $\widehat{W_x}$ estimation, and so does the GA at 1200 Hz. Regarding $\widehat{D_z}$, a bias in the order of 1 mm is observed with both the mono-frequency GAs. Such a result reveals the ill-posed nature of EC problems.
As far as the bi-frequency GA is concerned, accurate estimation of the 3 dimensions of the deeper crack is also carried out, the bias being close to zero for a 10 dB SNR.
These results go to prove that the proposed approach, consisting in a GA implementing bi-frequency EC images provided by the semi-analytic DPSM based model, is suitable for accurately reconstructing 3D surface cracks. Moreover, this method is rather computationally efficient, a drawing of such an algorithm requiring a computation time in the order of several minutes when implemented by means of Matlab™ on a 3.4 GHz PC.

Conclusion

In this paper, a method operating in bi-frequency applying to systems featuring homogeneous EC induction has been proposed for reconstructing 3D surface cracks. This method is an iterative approach relying on the use a semi-analytic direct model of EC images. With a view to enable the estimation of the depth of the cracks, bi-frequency EC images taking the skin effect into account have been used. The proposed method has been implemented in the form of a GA. Numerical experiments including comparison to mono-frequency GAs have been carried out for different SNR. The bi-frequency approach proved more efficient, and led to accurate estimation of the 3 dimensions of cracks even at low SNR.
Based on these results the 3D reconstruction of buried cracks can be envisaged. Therefore, further works will focus on a multi-frequency approach taking the skin effect into account. The application of the method to experimental data will also be considered.

References

[1] Kubinyi M, Docekal A, Ramos H and Ribeiro A 2010 *Przegląd Elektrotechniczny* **86** 249-254
[2] Bernieri A, Ferrigno L, Laracca M and Molinara M 2008 *Instrumentation and Measurement, IEEE Transactions on* **57**(9) 1958-1968
[3] Colton D and Päivärinta L 1992 *Archive for rational mechanics and analysis* **119**(1), 59-70
[4] Isakov, V. 1993 *Inverse Problems* **9**(6) 579
[5] Yamamoto M 1997 *International Journal of Applied Electromagnetics and Mechanics* **8**(1) 77-98
[6] Ribeiro A L, Ramos H G, Pasadas D and Rocha T 2014, February *40th Annual Review Of Progress In Quantitative Nondestructive Evaluation AIP Publishing* **1581**(1) 1428-1432
[7] Douvenot R, Lambert M and Lesselier D 2011 April Magnetics, *IEEE Transactions on* **47**(4) 746-755
[8] Cai C, Rodet T and Lambert M 2014 *Journal of Physics: Conference Series,* IOP Publishing **542**(1) 012009
[9] Tamburrino A, Calvano F, Ventre S. and Rubinacci G 2012 *NDT & E International* **47** 26-34.

[10] Tian, G. Y., Sophian, A. Taylor, D. and Rudlin, J 2005. *IEEE Sensors Journal*, 5(1), 90-96.
[11] Ribeiro, A L, Ramos, H G, & Arez, J C 2012. Measurement, 45(9), 2246-2253.
[12] Koyama K, Hoshikawa H and Taniyama N 2000, October *Proceedings of WCNDT*
[13] Thomas V, Joubert P Y, Vourc'h E and Placko D 2010, February *Sensors Applications Symposium (SAS IEEE)* 154-157
[14] Placko D, Bore T and Kundu T 2012 *Ultrasonic and Electromagnetic NDE for Structure and Material Characterization: Engineering and Biomedical Applications CRC Press* 249-294
[15] Vourc'h E, Bore T, Cai C and Soulat R 2015 *Journal of Physics: Conference Series*, 657(1), 012015.
[16] Goldberg D E and Holland J H 1988 *Machine Learning, Kluwer Academic Publishers* **3**(2-3), 95-99.
[17] Sabbagh H and Sabbagh L 1986 *Magnetics, IEEE Transactions on* **22**(4) 282-291

Institute of Physics
Dirac House, Temple Back
Bristol BS1 6BE UK

ISSN: 1742-6588
ISBN 978-1-5108-5035-4

CURRAN ASSOCIATES INC.
proceedings.com

6th International Workshop & Summer School on Plasma Physics 2014 (IWSSPP'14)

Journal of Physics: Conference Series Volume 768

Kiten, Bulgaria
30 June – 6 July 2014

IOP
Institute of Physics